Cambridge Primary
Revise for Primary Checkpoint
Mathematics

Second Edition

Paul Broadbent

Study Guide

HODDER
EDUCATION
AN HACHETTE UK COMPANY

This text has not been through the Cambridge International endorsement process.

Acknowledgements

The Publishers would like to thank the following for permission to reproduce copyright material. Every effort has been made to trace or contact all copyright holders, but if any have been inadvertently overlooked the Publishers will be pleased to make the necessary arrangements at the first opportunity.

Photo acknowledgements

p. 19 *cr* © Images My/Adobe Stock Photo; **p. 34** *cc* © Feng Yu/Adobe Stock Photo; **p. 35** *br* © Szasz-Fabian Erika/Adobe Stock Photo; **p. 44** *cr* © Yliv Design/Adobe Stock Photo; **p. 44** *cr* © Szasz-Fabian Erika/Adobe Stock Photo; **p. 50** *cr* © Gstudio/Adobe Stock Photo; **p. 52** *b* © Shiny 777/Adobe Stock Photo; **p. 52** *br* © BSvit/Adobe Stock Photo; **p. 57** *cr* © Zentangle/Adobe Stock Photo; **p. 57** *cr* © Ifh 85/Adobe Stock Photo; **p. 88** *br* © BNP Design Studio/Adobe Stock Photo; **p. 90** *tr* © Mast3r/Adobe Stock Photo; **p. 101** *b* © Zentangle/Adobe Stock Photo.

t = top, *b* = bottom, *l* = left, *r* = right, *c* = centre

Orders: please contact Hachette UK Distribution, Hely Hutchinson Centre, Milton Road, Didcot, Oxfordshire, OX11 7HH. Telephone: +44 (0)1235 827827. Email education@hachette.co.uk. Lines are open from 9 a.m. to 5 p.m., Monday to Saturday, with a 24-hour message answering service. You can also order through our website: www.hoddereducation.com

© Paul Broadbent 2022

First edition published in 2013

This edition first published in 2022 by

Hodder Education
An Hachette UK Company
Carmelite House
50 Victoria Embankment
London EC4Y 0DZ

www.hoddereducation.com

Impression number   10 9 8 7 6 5 4 3 2 1
Year   2026 2025 2024 2023 2022

Cover artwork by Lisa Hunt, The Bright Agency

Illustrations by Natalie and Tamsin Hinrichsen, Stéphan Theron

Typeset in FS Albert 12/14 by IO Publishing CC

Printed in the UK

A catalogue record for this title is available from the British Library.

ISBN: 9781398369856

Contents

What is this book about?

This Study Guide is to help you revise and practice important skills and concepts you have learnt in preparation for the Cambridge Primary Checkpoint Mathematics test. It will help you to recall key information and ideas and build your understanding about the maths topics that you have been learning during Stage 6.

The book is divided into three chapters: Number, Geometry and Measure, and Statistics and Probability. At the end of each chapter there is a 'Test your Understanding' set of questions to help you check your progress. Each chapter has a number of units which are broken down into the key topics for each unit.

Do you remember?

These are the most important pieces of information you need to know about a topic. It includes methods, facts and explanations to help you, with examples (in white boxes) for you to work through.

Maths words

These words relate to the area of maths that you are revising.

It is important to understand what they mean, so some of the words are also included in a glossary on pages 111–112.

Practise

Each question will help you to practise the mathematical skills and methods you need to know. Most recording of answers is on the page, with some problems and questions needing paper for the working out.

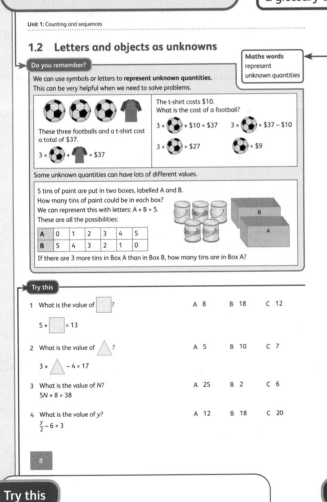

Try this

This has four multiple choice questions related to the 'Do you remember?' section. It is a quick warm-up and also checks you have understood the topic being covered.

Thinking mathematically

For these more challenging activities, you will need to use your reasoning and problem solving skills. They often involve several steps to solve them and you may need to use extra paper to record your working.

1.1 Number sequences

Do you remember?

A number **sequence** is a list of numbers that follow a **pattern**.
Each number in the sequence is called a **term**.
You can often find the pattern or rule to work out the next
term in a sequence by looking at the **difference** between the numbers.

These sequences include **negative numbers**. What is the next number in each of them?

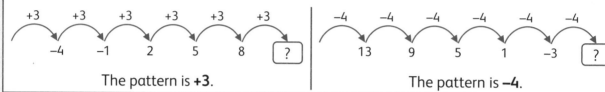

The pattern is **+3**.   The pattern is **−4**.

This sequence includes decimal fractions.
Try counting on and back to work out the missing number.

0.35   0.7   1.05   [?]   1.75   2.1

Try this

1   What is the missing number in this sequence?
    A 4   B 3   C 5

[−9], [−5], [−1], [   ], [7], [11]

2   What is the next number in this sequence?
    A 16.6   B 16.1   C 17.1

[19.6], [19.1], [18.6], [18.1], [17.6], [   ]

3   What is the missing number in this sequence?
    A 4   B 6   C 5

[−27], [−16], [−5], [   ], [17], [28]

4   What is the next number in this sequence?
    A $\frac{1}{2}$ B $-\frac{1}{2}$ C $-\frac{3}{4}$

[$\frac{3}{4}$], [$\frac{1}{2}$], [$\frac{1}{4}$], [0], [$-\frac{1}{4}$], [   ]

Practise

1   Write the pattern or rule for each sequence.

    a   −6   −1   4   9   14   The rule is _____.

    b   20   14   8   2   −4   The rule is _____.

    c   11   7   3   −1   −5   The rule is _____.

    d   −52   −22   8   38   68   The rule is _____.

    e   1.2   0.7   0.2   −0.3   −0.8   The rule is _____.

2 Write the missing numbers in each sequence.

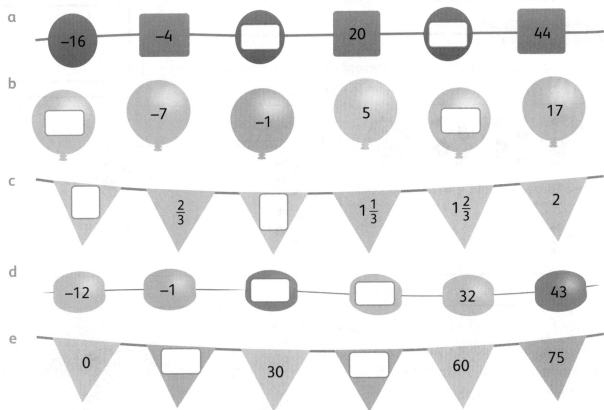

a

−16   −4   ☐   20   ☐   44

b

☐   −7   −1   5   ☐   17

c

☐ $\frac{2}{3}$ ☐ $1\frac{1}{3}$ $1\frac{2}{3}$ 2

d

−12   −1   ☐   ☐   32   43

e

0   ☐   30   ☐   60   75

3 Write the tenth term for each of these sequences.

a

Position	Term
1	2
2	4
3	6
4	8
10	

b

Position	Term
1	5
2	10
3	15
4	20
10	

c

Position	Term
1	9
2	18
3	27
4	36
10	

Thinking mathematically

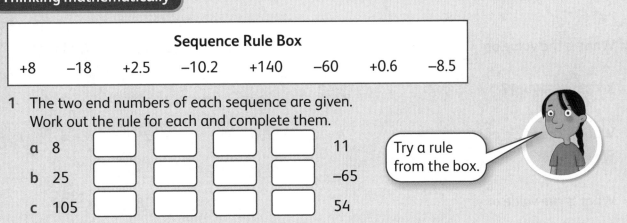

Sequence Rule Box

+8   −18   +2.5   −10.2   +140   −60   +0.6   −8.5

1 The two end numbers of each sequence are given.
Work out the rule for each and complete them.

a 8   ☐ ☐ ☐ ☐   11

b 25   ☐ ☐ ☐ ☐   −65

c 105   ☐ ☐ ☐ ☐   54

Try a rule from the box.

2 Make up your own sequences using rules from the Sequence Rule Box.

1.2 Letters and objects as unknowns

Maths words

represent

unknown quantities

Do you remember?

We can use symbols or letters to **represent unknown quantities.**
This can be very helpful when we need to solve problems.

These three footballs and a t-shirt cost a total of $37.

$3 \times$  $+$ 👕 $= \$37$

The t-shirt costs $10.
What is the cost of a football?

$3 \times$ ⚽ $+ \$10 = \$37$ $3 \times$ ⚽ $= \$37 - \$10$

$3 \times$ ⚽ $= \$27$ ⚽ $= \$9$

Some unknown quantities can have lots of different values.

5 tins of paint are put in two boxes, labelled A and B.
How many tins of paint could be in each box?
We can represent this with letters: A + B = 5.
These are all the possibilities:

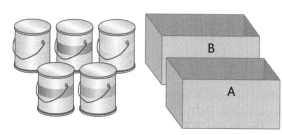

A	0	1	2	3	4	5
B	5	4	3	2	1	0

If there are 3 more tins in Box A than in Box B, how many tins are in Box A?

Try this

1 What is the value of ☐ ?

    $5 +$ ☐ $= 13$

A 8   B 18   C 12

2 What is the value of △ ?

    $3 \times$ △ $- 4 = 17$

A 5   B 10   C 7

3 What is the value of N?
    $5N + 8 = 38$

A 25   B 2   C 6

4 What is the value of y?
    $\frac{y}{2} - 6 = 3$

A 12   B 18   C 20

Practise

1 Write the answers for each of these, when:

(w = 6)   (x = 8)   (y = 4)   (z = 7)

a $6w$ → _____   b $w + z$ → _____

c $2x + z$ → _____   d $3y - w$ → _____

2 A pen and ruler cost $10 in total. The pen costs $4 more than the ruler.
What is the price of each? Show this as an equation, with pen = p and ruler = r.

_____

_____

3 The letters c and d stand for two whole numbers.

($c \times d = 24$)

a Which two numbers could c and d represent?

c							
d							

b What if $c \times d = 24$ and $c - d = 5$?
Look at your answers in the table above to find c and d.

$c =$ ☐ $d =$ ☐

c Try this. What is the value of c and d?
$c \times d = 30$ and $c - d = 7$

c							
d							

$c =$ ☐ $d =$ ☐

Thinking mathematically

1 The formula for finding the area of a rectangle is **Area = length × width**.
This can be written as $A = l \times w$.

2.5 cm   Area = 20 cm^2

$l =$ ☐ ?

Use this formula to find the missing length on this rectangle: $l =$ ☐ cm

2 The formula for finding the perimeter of a rectangle is:
Perimeter = 2(length + width)
This can be written as $P = 2(l + w)$.
Use this formula to find the perimeter of this rectangle: $P =$ ☐ cm

2.1 Addition and subtraction

Do you remember?

When you **add** and **subtract**, **estimate** an **approximate** answer first.

To find an approximate answer choose to **round** to the nearest 10 or 1 to make the numbers easy to calculate in your head.

What is 364.74 added to 107.49?
An approximate answer is 360 + 110 = 470

```
    3  6 ¹4 . ¹7 4
  + 1  0  7 .  4 9
    4  7  2 .  2 3
```

What is 4.651 subtract 1.965?
An approximate answer is 5 − 2 = 3

```
    ³4́ . ¹5₆ ¹4₅ ¹1
  − 1  .  9  6  5
    2  .  6  8  6
```

We can use a number line to help us calculate with **negative numbers**.

What is the **difference** between −4 and 5?

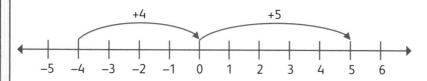

The difference between −4 and 5 is 9.   −4 + 9 = 5
                                           5 − 9 = −4

Maths words
add
subtract
estimate
approximate
round
negative numbers
difference

Try this

1   What is 295.85 subtract 87.68?
    A   206.27
    B   208.17
    C   212.23

2   What is the total of $321.49 and $653.18?
    A $974.67
    B $974.57
    C $874.67

3   Two suitcases weigh 29.47 kg and 54.28 kg.
    What is the difference in their weight?
    A   35.21 kg
    B   25.71 kg
    C   24.81 kg

4   What is the sum of 491.83 and 158.78?
    A   650.51
    B   649.61
    C   650.61

Practise

1 Read and answer these problems.

   a What is the sum of 235.88 and 129.26? _____

   b What is the total of 1.717 and 4.355? _____

   c What is the difference between 56.18 and 26.35? _____

   d What is 700.63 subtract 291.44? _____

2 Answer these questions.

A

3.245 kg

B

11.88 kg

C

2.915 kg

D

4.203 kg

   a What is the total weight of parcels C and D? ▭

   b What is the difference in weight between parcels A and C? ▭

   c How much more does parcel D weigh than parcel A? ▭

   d What is the total weight of parcels A, B and D? ▭

3 The digits 1, 2 and 3 are missing from each of these. Write the completed calculations.

   a
   ```
       4 □ . □ 5
   +   5 0 . 9 □
   ─────────────
       9 4 . 0 7
   ```

   b
   ```
       8 4 . 0 □
   -   □ 9 . □ 5
   ─────────────
       6 4 . 7 8
   ```

4 This table shows the minimum temperature in a city during one week in January.

Mon 10th	Tues 11th	Wed 12th	Thurs 13th	Fri 14th	Sat 15th	Sun 16th
0 °C	−3 °C	−7 °C	−4 °C	3 °C	5 °C	9 °C

   a What is the difference between the highest and lowest temperatures this week? ▭ °C

   b What is the increase in temperature from Thursday to Friday? ▭ °C

Thinking mathematically

Look at this grid of numbers.

5.093	9.839	7.283	3.281
9.719	1.307	9.473	8.161
6.527	5.717	2.907	4.693

Match the pairs of numbers that, when added, will give whole number totals.

2.2 Mental calculation strategies

Do you remember?

There are many ways to **add** or **subtract** numbers in your head.

- Use any facts you know to help learn others.

Maths words
add
subtract
partition

8 + 6 = 14

You can use this to work out these and other facts:

18 + 16 = 34   80 + 60 = 140   1.8 + 1.6 = 3.4

- **Partition** (break up) numbers to make them easier to work with.

3.6 + 5.7 = ___
- Hold the bigger number in your head (5.7)
- Add on the ones (5.7 + 3 = 8.7)
- Add on the tenths (8.7 + 0.6 = 9.3)

So, 3.6 + 5.7 = 9.3

$$71 - 46 = (60 + 11) - (40 + 6)$$
$$= (60 - 40) + (11 - 6)$$
$$= 20 + 5$$
$$= 25$$

- Count on from the smallest number to work out differences.

What is the difference between 170 and 320?
Count on from 170 to 200 and then to 320.

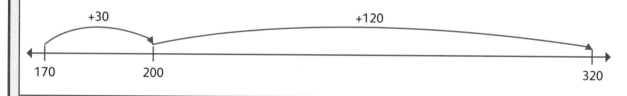

+30   +120

170   200   320

Try this

1 What is 48 add 35?   A 73   B 83   C 93

2 What is 2.3 + 1.9?   A 3.2   B 4.2   C 4.4

3 What is the difference between 145 and 129?   A 16   B 14   C 24

4 What is the missing number?   ☐ − 18 = 24   A 34   B 6   C 42

Practise

1 Use the first answer to help with the others.

a 7 + 6 = ☐

b 70 + 60 = ☐   c 700 + 600 = ☐   d 17 + 16 = ☐

e 1007 + 3006 = ☐   f 2070 + 4060 = ☐   g 0.7 + 0.6 = ☐

2 Answer these questions.

   a What is the sum of 64 and 87? _____

   b What is the difference between 83 and 58? _____

   c What is the total of 14, 15 and 16? _____

   d What is 74 subtract 36? _____

3 Complete this addition square.

+	9	13	30
5			35
43		56	
80	89		

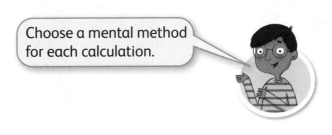

Choose a mental method for each calculation.

4 Look at the number machines. Complete each table.

   a IN → +19 → OUT

IN	8		14		12		18	
OUT		25		32		30		38

   b IN → +29 → OUT

IN	7		16		11		15	
OUT		37		32		41		47

Thinking mathematically

You need a set of digit cards 1–9.
Shuffle and take any four cards.
Make two 2-digit numbers.

- Find the total of the two numbers.
- Find the difference between the two numbers.
- Rearrange the four digits.
  What is the largest total and smallest difference you can make?

2.3 Brackets and order of operations

Do you remember?

Calculations should be carried out using the following **order** of **operations**.

- **Brackets:** work out any brackets first
- Division
- Multiplication
- Addition
- Subtraction

Maths words

order

operations

brackets

$5 × (3 + 8) − 2$
$= 5 × 11 − 2$
$= 55 − 2$
$= 53$

$(6 × 2 − 8) + (12 ÷ 4 + 10)$
$= (12 − 8) + (3 + 10)$
$= 4 + 13$
$= 17$

Try this

1 What is the missing number?

$\boxed{} ÷ 8 + 3 = 8$

A 40   B 32   C 88

2 What is the missing number?

$3 × \boxed{} − 15 = 9$

A 3   B 9   C 8

3 What is the missing number?

$(19 − 7) + \boxed{} = 18$

A 1   B 6   C 8

4 What is the missing number?

$4 × (9 − \boxed{}) = 24$

A 12   B 6   C 3

Practise

1 Answer each of these number sentences. Remember to first calculate the numbers within brackets.

a $(13 − 5) × 2 =$ $\boxed{}$

b $3 × (8 − 5) =$ $\boxed{}$

c $(4 + 6) ÷ 2 =$ $\boxed{}$

d $(8 + 2) − (3 + 5) =$ $\boxed{}$

e $(9 × 2) + (4 × 5) =$ $\boxed{}$

f $(15 − 9) + (13 − 7) =$ $\boxed{}$

2 Draw brackets to make each number sentence answer 12.

a $19 − 12 − 5$

b $16 − 10 − 6$

c $22 − 5 + 5$

d $6 + 13 − 7$

e $24 − 6 − 6$

f $20 − 10 − 2$

3 Write the missing numbers.

a (⬚ × 4) − 1 = 11

b 10 − (⬚ × 3) = 4

c (4 × 2) + (⬚ × 3) = 17

d (⬚ × 5) − (5 × 4) = 10

e 12 ÷ (⬚ × 2) = 2

f (⬚ × 5) ÷ 2 = 10

4 Put brackets in these calculations to make different answers.
Record the different answers you can make.

a 19 − 10 − 5 − 2 = ⬚

b 25 − 4 + 9 × 2 = ⬚

c 15 + 11 − 9 − 6 = ⬚

d 6 × 3 + 5 − 4 = ⬚

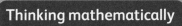

Thinking mathematically

Work out the mystery number for each of these.

What's my number?

a When I double my number and then add 3, the answer is 17. ⬚

b When I divide my number by 3 and then add 5, the answer is 12. ⬚

c When I multiply my number by 5 and then subtract 6, the answer is 39. ⬚

d When I divide my number by 4 and then subtract 2, the answer is 3. ⬚

2.4 Multiplication

Maths words

multiply

multiples

estimate

Do you remember?

To **multiply** with large numbers, you need to be able to multiply **multiples** of 10.

$34 \times 26 = \boxed{}$

Look at these two methods:

Grid method:

×	30	4	
20	600	80	→ 680
6	180	24	→ + 204

884

Vertical method:

```
        3  4   → leading to →        3  4
  ×     2  6                    ×     2  6
     6  0  0   (30 × 20)           6  8  0  (34 × 20)
        8  0   (4 × 20)            2  0  4  (34 × 6)
     1  8  0   (30 × 6)            8  8  4
        2  4   (6 × 4)
     8  8  4
```

It is important to **estimate** an answer before multiplying.

584×46

Estimate:

$\approx 600 \times 50$

$\approx 30\,000$

```
         5  8  4
  ×         4  6
      3  5  0  4   (584 × 6)
   2  3  3  6  0   (584 × 40)
   2  6  8  6  4
```

Compare your estimate answer with the final answer.

Try this

1 What is the missing number?

$70 \times \boxed{} = 2100$

A 3   B 30   C 300

2 What is 1 500 multiplied by 6?

A 9 000   B 6 500   C 8 000

3 What is the missing number?

$\boxed{} \times 42 = 1680$

A 44   B 45   C 40

4 What is 38 multiplied by 26?

A 648   B 988   C 1 048

Practise

1   Use the grid method to complete these multiplications.

   a   84 × 37 = [   ]

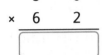

×	80	4
30		
7		

   → [   ]
   → + [   ]
   [   ]

   b   62 × 43 = [   ]

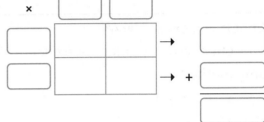

   → [   ]
   → + [   ]
   [   ]

2   Estimate then answer these.

   a
   ```
        5   6
   ×    6   2
   ```
   [   ]

   b
   ```
        2   9
   ×    8   4
   ```
   [   ]

   c
   ```
        2   0   3
   ×        7   4
   ```
   [   ]

   d
   ```
        8   4   5
   ×        1   6
   ```
   [   ]

3   Estimate then answer these.

   a
   ```
        5   5   6   7
   ×                8
   ```
   [   ]

   b
   ```
        8   5   6   1
   ×                6
   ```
   [   ]

   c
   ```
        6   5   4   2
   ×            1   7
   ```
   [   ]

   d
   ```
        8   5   6   4
   ×            2   6
   ```
   [   ]

4   Answer these problems.

   a   A cinema has 46 rows with 38 seats in each row.
       How many seats are there in total?

   b   Each side of a square tile is 76 cm. What is the area of the tile?

   c   A ticket for a flight from Cairo to London costs $487. What is the total cost of
       tickets for a group of 28 travellers?

   d   A chocolate bar weighs 186 g. How much would a box of 18 bars of chocolate
       weigh, including the box which weighs 145 g?

[   ]
[   ]
[   ]
[   ]

Thinking mathematically

Explore this multiplication pattern.

Multiply 15 873 by 7.
Now multiply 15 873 by any multiple of 7.
What patterns do you notice?

```
     15 873
×         7
[        ]
```

```
     15 873
×        14
[        ]
```

2.5 Division

Maths words
divide
remainder
quotient

Do you remember?

If a number cannot be **divided** exactly, it leaves a **remainder**.

What is 749 divided by 4?

Work out how many groups of 4 are in 749 and what is left over:

Long method:

```
      1  8  7  r 1
4 ) 7  4  9
   -4  0  0    (4 × 100)
    3  4  9
   -3  2  0    (4 × 80)
       2  9
      -2  8    (4 × 7)
          1
```

Short method:

```
      1  8  7  r 1
4 ) 7  ³4  ²9
```

749 ÷ 4 = 187 remainder 1

We can turn the remainder into a fraction.

1 out of 4 gives $\frac{1}{4}$, so the answer is $187\frac{1}{4}$.

Remember to estimate first and to align digits in the correct columns.

You can check your answers to division calculations (the **quotient**) using multiplication.

Try this

1 What is the missing number?

   $216 \div \boxed{} = 30\,r\,6$

   A 7 B 8 C 6

2 What is the remainder when 345 is divided by 6?

   A 1 B 2 C 3

3 What is the missing number?

   $\boxed{} \div 9 = 45$

   A 375 B 5 C 405

4 How many 3 metre lengths of thread can be cut from a reel of 136 metres?

   A 46 B 45 C 44

Practise

1 Answer these using the long or short method.

   a 9)8 461   b 6)1 196   c 5)3 408   d 7)5 067

2 Complete these. Write the remainders as fractions.

a 359 ÷ 10 = ☐

b 856 ÷ 3 = ☐

c 764 ÷ 9 = ☐

d 541 ÷ 25 = ☐

3 Read and answer these problems.

a There are 366 days in a leap year.
How many full weeks are there and how many days are left over?

☐ weeks and ☐ days

b How many 5 cm lengths can be cut from 133 cm of string?

☐ lengths

How much string is left over?   ☐ cm

c 138 flowers are divided into bunches of nine flowers.
How many bunches of flowers can be made?

☐ bunches

How many flowers are left over?   ☐ flowers

4 Use these numbers to answer the problems.

| 116 | 667 | 780 | 425 | 762 |

a Which numbers are exactly divisible by 5?   _____

b Which of the numbers has a remainder of 1 when the divisor is 4?   ☐

c Which of the numbers can be divided exactly by 6?   _____

d Which of the numbers has a remainder of 2 when the divisor is 3?   _____

e Which number is exactly divisible by 25?   ☐

5 Sam has a collection of badges that he wants to count.
He knows that he has between 120 and 150 badges, but not the exact number.
He decides to count them in fives, and he has 2 left over.
He then counts them in sixes and he has 3 left over.
How many badges does Sam have?   ☐

Thinking mathematically

a Find three numbers between 100 and 140 that have:

• a remainder of 1 when divided by 2   _____

• a remainder of 3 when divided by 4   _____

• a remainder of 5 when divided by 6   _____

b Can you predict the next two numbers after 140 that will have the same remainders?
Check your prediction by dividing.   _____

2.6   Factors and primes

Do you remember?

Factors are whole numbers that will divide exactly into other whole numbers.

> 24 can be divided exactly by 1, 2, 3, 4, 6, 8, 12 and 24.
> So, the factors of 24 are 1, 2, 3, 4, 6, 8, 12 and 24.

A **common factor** is a number which is a factor of two or more numbers.

> To find the common factors of 8, 24 and 32:
> The factors of 8 are 1, 2, 4 and 8.
> The factors of 24 are 1, 2, 3, 4, 6, 8, 12 and 24.
> The factors of 32 are 1, 2, 4, 8, 16 and 32.
> The common factors of 8, 24 and 32 are 1, 2, 4 and 8.
> The **highest common factor** (HCF) is 8.

If a number only has 2 factors, itself and 1, then it is a **prime number**.

> 17 is a prime number because it can only be divided exactly by 1 and 17.

A **prime factor** is simply a factor that is a prime number.

> The factors of 12 are 1, 2, 3, 4, 6 and 12.
> The prime factors are 2 and 3.

Maths words

factor
common factor
highest common factor
prime number
prime factor

Try this

1   Which of these numbers is a factor of 63?
    A 8   B 6   C 9

2   Which of these numbers is a common factor of 18 and 60?
    A 6   B 10   C 9

3   Which of these numbers is the highest common factor of 28 and 35?
    A 5   B 7   C 9

4   Which of these numbers is the highest common factor of 60 and 96?
    A 6   B 3   C 12

Practise

1   Write the factors of these numbers in order, starting with 1.

    a   45 → 1 ☐ ☐ ☐ ☐ ☐

    b   36 → 1 ☐ ☐ ☐ ☐ ☐ ☐ ☐ ☐

    c   30 → 1 ☐ ☐ ☐ ☐ ☐

    d   42 → 1 ☐ ☐ ☐ ☐ ☐ ☐ ☐

2   What are the common factors for each of these numbers?

   a   27 and 45   →   _____

   b   14 and 42   →   _____

   c   36 and 48   →   _____

   d   12, 14 and 30 →   _____

   e   27, 30 and 42 →   _____

Circle the highest common factor (HCF) in each of your answers.

3   What are the prime factors of these numbers

   a   42   b   90   c   72   d   132

   _____   _____   _____   _____

4   Write the numbers from 1 to 20 in the correct place on this Venn diagram.

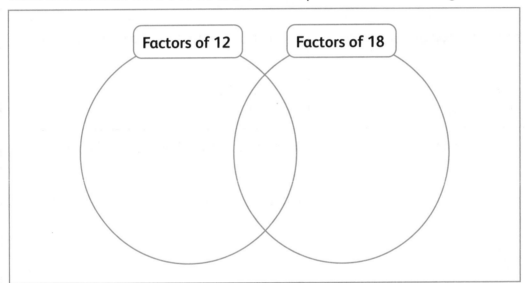

Factors of 12   Factors of 18

Thinking mathematically

Write the following numbers on the grid to make it true.

	Prime factor	Common factor of 18 and 12	Factor of 72
Factor of 60			
Odd number			
Factor of 24			

1   2   3

4   5   6

7   8   9

2.7 Tests of divisibility

Do you remember?

Tests of divisibility are a very useful way of quickly working out if a number can be exactly **divided** by another number. Use your knowledge of **multiples** to help you.

A whole number is divisible by, or is a multiple of:

Maths words
divided
multiple

2 if the last digit is even. Examples: 36, 194, 2 116	**6** if it is even and the sum of its digits is divisible by 3. Examples: 528 (5 + 2 + 8 = **15**) 402 (4 + 0 + 2 = **6**)
3 if the sum of its digits can be divided by 3. Examples: 135 (1 + 3 + 5 = **9**) 2 193 (2 + 1 + 9 + 3 = **15**)	**8** if half of the number is divisible by 4. Examples: 264 (÷ 2 = **132**) 432 (÷ 2 = **216**)
4 if the last two digits can be divided by 4. Examples: 1**24**, 3**40**, 2 5**64**	**9** if the sum of its digits is divisible by 9. Examples: 135 (1 + 3 + 5 = 9) 8 208 (8 + 2 + 0 + 8 = 18)
5 if the last digit is 0 or 5. Examples: 32**0**, 14**5**, 3 02**5**	**10** if the last digit is 0. Examples: 29**0**, 35**0**, 68**0**, 41**0**

Try this

1 Which of these numbers is divisible by 9?
   A 217   B 3 904   C 1 737

2 Which of these numbers is divisible by 4?
   A 199   B 208   C 334

3 Which of these numbers is a common multiple of both 3 and 4?
   A 288   B 375   C 416

4 Which of these numbers is a common multiple of both 5 and 9?
   A 380   B 621   C 495

Practise

1   Write the numbers 2, 3, 4, 5, 6, 8 or 9 in the correct boxes.
    Use the tests of divisibility to find the answers.

    a   1 468 is divisible by ☐ and ☐.

    b   2 745 is divisible by ☐, ☐ and ☐.

    c   6 102 is divisible by ☐, ☐, ☐ and ☐.

    d   24 096 is divisible by ☐, ☐, ☐, ☐ and ☐.

2   Use these numbers to answer each question.

| 860 | 591 | 96 | 85 | 450 | 72 |

    a   Which numbers are multiples of 4?   _____
    b   Which numbers are multiples of 6?   _____
    c   Which numbers are multiples of 9?   _____
    d   Which numbers are common multiples of 3 and 4?   _____
    e   Which numbers are common multiples of 2, 5 and 10?   _____
    f   Which numbers are common multiples of 3, 6, and 9?   _____

3   Investigate divisibility of any number by 2, 3, 4, 5, 6, 8, 9 or 10:

    a   What is the smallest number that is divisible by exactly 3 numbers?   _____
    b   What is the smallest number that is divisible by exactly 4 numbers?   _____
    c   What is the smallest number that is divisible by exactly 5 numbers?   _____
    d   What is the smallest number that is divisible by exactly 6 numbers?   _____
    e   What do you notice about all these multiples? _____

4   Write the numbers 20 to 40 in the correct places on this Carroll diagram.

	Multiples of 4	Not multiples of 4
Divisible by 6		
Not divisible by 6		

Thinking mathematically

Sanchia thinks her statement is always true.

Test what she says using some sets of 2-digit and 3-digit consecutive numbers.

Is she always, sometimes or never true?

What do you notice about products of 3 consecutive numbers?

The product of any 3 consecutive whole numbers is divisible by 6.

23

2.8 Square, triangular and cube numbers

Do you remember?

The numbers 1, 4, 9 and 16 are examples of **square numbers**, shown as a pattern of squares.

   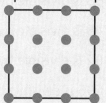

$1 \times 1 = 1$	$2 \times 2 = 4$	$3 \times 3 = 9$	$4 \times 4 = 16$
1 squared = 1	2 squared = 4	3 squared = 9	4 squared = 16
$1^2 = 1$	$2^2 = 4$	$3^2 = 9$	$4^2 = 16$

Triangular numbers can also be shown as patterns.

  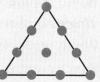

Position	1	2	3	4
Term	1	3	6	10

A square number is the product of two equal numbers.
We use the special symbol ^{2} to mean squared.

6

$6 \times 6 = 36$ 6 squared is 36 $6^2 = 36$

6

A **cube number** is the product of three equal numbers.
We use the special symbol ^{3} to mean cubed.

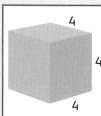
4

4   $4 \times 4 \times 4 = 64$ 4 cubed is 64 $4^3 = 64$

4

Maths words
square number
triangular number
term
cube number

Try this

		A	B	C
1	What is 13^2?	A 26	B 159	C 169
2	What is 5^3?	A 45	B 125	C 150
3	What is the fifth triangular number?	A 18	B 16	C 15
4	What is the 12th square number?	A 121	B 144	C 240

Practise

1 Continue each sequence.

a

Position	1	2	3	4	5	6	7	8	9	10
Term	1	4	9	16						

b

Position	1	2	3	4	5	6	7	8	9	10
Term	1	3	6	10						

Look at the differences between each term. Describe any patterns you notice.

2 Write these numbers in full.

a $11^2 =$ ☐   b $1^3 =$ ☐   c $6^3 =$ ☐   d $20^2 =$ ☐

e $5^3 =$ ☐   f $10^3 =$ ☐   g $30^2 =$ ☐   h $12^3 =$ ☐

3 Write the numbers 1–40 on this Venn diagram.

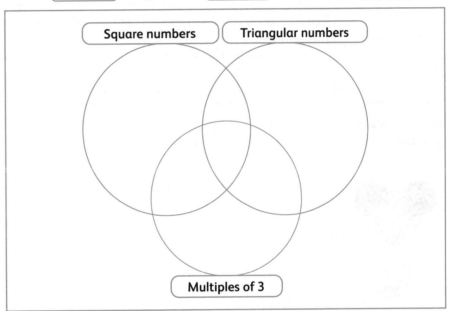

Square numbers   Triangular numbers

Multiples of 3

Thinking mathematically

32 is a Happy Number!

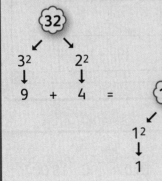

Find other numbers to 50 that are Happy Numbers.

This is how to work out if a number is happy.

a Choose a number.

b Square the digits and add.

c Repeat this for each new number.

If you end with a 1, the number you started with is happy.

3.1 Place value

Do you remember?

Numerals are made up of ten **digits** from 0–9.
Each digit of a number has a **place value**.
Look at the position of each digit to work out its value.

Maths words
numeral
digit
place value
decompose
decimal point

Say this number aloud:
583 136
five hundred and eighty-three thousand, one hundred and thirty-six.

Hundred thousands	Ten thousands	Thousands	Hundreds	Tens	Ones
5	8	3	1	3	6

To **decompose** a number, break it up to show the value of each digit:
500 000 + 80 000 + 3 000 + 100 + 30 + 6 = 583 136

With decimal numbers, the **decimal point** separates whole numbers from decimal fractions.

18.315 is read as **eighteen point three one five**.

Tens	Ones		Tenths	Hundredths	Thousandths
1	8	.	3	1	5
(10)	(8)		$(\frac{3}{10})$	$(\frac{1}{100})$	$(\frac{5}{1000})$

Try this

1 What is $15\frac{7}{100}$ as a decimal?   A 15.7   B 15.07   C 15.007

2 What is the value of the red digit in 698 304?   A 30   B 3 000   C 300

3 Which number is six point one two nine?   A 6.219   B 6.29   C 6.129

4 What is 0.781 as a fraction?   A $7\frac{81}{100}$ B $\frac{781}{1000}$ C $78\frac{1}{10}$

Practise

1 Read these and write each as a numeral.

   a three hundred thousand, nine hundred and twenty-five _____

   b eighty-nine thousand, four hundred and seventy-nine _____

   c four hundred and seventeen thousand, five hundred and sixty-one _____

   d two hundred and one thousand, three hundred and ninety _____

2 Decompose these numbers to show the value of each digit.

   a 34 870   b 96 105   c 448 033   d 700 823

3 Write the decimal number each arrow points to.

   a 18   18.5   19

   [ ]   [ ]   [ ]   [ ]   [ ]

   b 0.6   0.65   0.7

   [ ]   [ ]   [ ]   [ ]   [ ]

4 Write the value of the red digit as a fraction.

   a 0.473 [ ]   b 5.981 [ ]   c 62.359 [ ]

   d 0.702 [ ]   e 44.008 [ ]   f 96.28 [ ]

5 Decompose these decimal numbers to show the value of each digit.

   a 38.65   b 9.055   c 34.583   d 51.608

Thinking mathematically

Use these four digits and a decimal point to answer these.
There must be one digit in front of the decimal point.

[4]   [8]   [9]   [1]   [.]

a What is the largest decimal number you can make?

b What the smallest decimal number you can make?

c Make a decimal number as near as possible to 4.

d Make a decimal number as near as possible to 9.

e Make a decimal number between 1.5 and 1.9.

[ ] . [ ] [ ] [ ]
[ ] . [ ] [ ] [ ]
[ ] . [ ] [ ] [ ]
[ ] . [ ] [ ] [ ]
[ ] . [ ] [ ] [ ]

3.2 Multiplying and dividing by 10, 100 and 1000

Do you remember?

When you **multiply** and **divide** by 10 or 100, keep the **decimal point** lined up and move the **digits**.

To multiply any number by 10, move the digits **one place to the left** and fill the space with a zero. ×10 3.64 36.40	To divide any number by 10, move the digits **one place to the right**: ÷10 14.83 1.483
To multiply any number by 100, move the digits **two places to the left** and fill the spaces with zeros: ×100 8.095 809.500	To divide any number by 100, Move the digits **two places to the right**: ÷100 394.7 3.947

Maths words
multiply
divide
decimal point
digits
zero

Remember ... putting a **zero** on the end of a decimal doesn't change the number. 1.2 is the same as 1.20 and 1.200.

Try this

1   What is 3.057 × 10?   A 13.057   B 3.57   C 30.57

2   What is 348 ÷ 100?   A 34 800   B 3.48   C 34.8

3   What is the missing number? ☐ ÷ 10 = 14.3   A 1.43   B 143   C 1 430

4   What is the missing number? ☐ × 100 = 28.6   A 0.286   B 2 860   C 2.86

Practise

1   Complete these.

    a   Multiply these numbers by 10.

        4.6   → ☐

        5.17 → ☐

        10.9 → ☐

        22.08 → ☐

    b   Multiply these numbers by 100.

        3.3   → ☐

        67.4 → ☐

        8.08 → ☐

        49.52 → ☐

c   Divide these by 10.

6.1   →   ☐

13.5   →   ☐

9   →   ☐

74.01   →   ☐

d   Divide these by 100.

345   →   ☐

267.7   →   ☐

99   →   ☐

8   →   ☐

2   Write the missing numbers in each of these.

a   ☐ × 10 = 39

b   6.23 × ☐ = 623

c   ☐ × 100 = 8 577

d   0.81 × ☐ = 8.1

e   ☐ ÷ 100 = 1.944

f   9 261 ÷ ☐ = 92.61

g   ☐ ÷ 10 = 17.3

h   885.7 ÷ ☐ = 88.57

3   Read and answer these.

a   A bucket holds 4.15 litres of water. How much water is there in 10 buckets?   ☐

b   A 250 kg sack of rice is divided into 100 packs. How much does each pack weigh?   ☐

c   A running track is 0.4 km long. A runner runs 10 laps. How many km has she run?   ☐

d   A can holds 0.33 litres of juice. How much juice is there in 100 cans?   ☐

4   Write the missing numbers going in and out of these multiplication machines.

a
3.4   IN
1.25
☐
☐
90.3

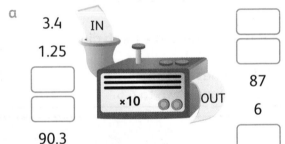

×10   OUT

☐
☐
87
6
☐

b
2.68   IN
9.4
☐
☐
0.18

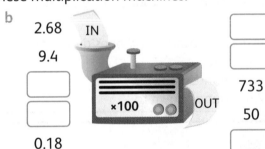

×100   OUT

☐
☐
733
50
☐

Thinking mathematically

This is a **super-square**.

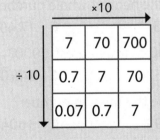

×10

7	70	700
0.7	7	70
0.07	0.7	7

÷ 10

Make up some of your own **super-squares**.

Complete these **super-squares**.

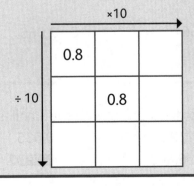

×10

3.2		
	0.32	

÷ 10

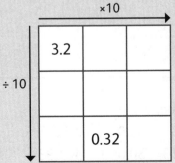

×10

0.8		
	0.8	

÷ 10

3.3 Comparing, ordering and rounding numbers

Do you remember?

The symbols > and < are used to **compare** numbers.

< means 'is less than'   3.85 < 4.05

> means 'is greater than'   52.2 > 49.8

If you have a list of numbers to put in **order**, look carefully at the value of each **digit**.

Write them in a column, lining up each number so they can be compared.

Positive numbers are above zero and negative numbers are below zero.

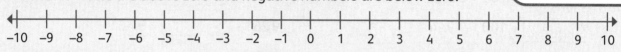

We can compare numbers by looking at their positions on the number line.

We **round** numbers to make them easier to work with. It is useful for **estimating approximate**, or rough, answers.

Whole numbers can be rounded to the nearest 10, 100 or 1 000. Decimal numbers can be rounded to the nearest whole number or tenth.

Rounding to the nearest whole number	Rounding to the nearest tenth
• Look at the tenths digit • If it is 5 or more, round up to the next whole number • If it is less than 5, the ones digit stays the same: 16.5 rounds up to 17 7.48 rounds down to 7.	• Look at the hundredths digit • If it is 5 or more, round up to the next tenth • If it is less than 5, the tenth digit stays the same: 13.77 rounds up to 13.8 4.639 rounds down to 4.6.

Maths words
compare
order
digit
round
estimating
approximate

Try this

1 Which number rounds to 17 as the nearest whole number?

   A 16.47   B 17.73   C 17.46

2 Which number rounds to 245 000 as the nearest 1 000?

   A 245 625   B 244 534   C 244 099

3 Which of these numbers could be the missing number?   3 554 < ☐

   A 3 468   B 3 099   C 3 604

4 Which number is the largest?

   A 2.606   B 2.66   C 2.066

Practise

1 Write each set of numbers in order, starting with the smallest number.

   a 19.407   19.74   19.007   19.9   b 0.035   0.033   0.302   0.32

   c 6.445   6.534   6.359   6.442   d 30.932   30.913   30.093   30.193

2   Complete this table.

	a Round to the nearest 100	b Round to the nearest 1 000	c Round to the nearest 10 000
892 388 →			
372 105 →			
1 093 465 →			
1 435 476 →			

3   Round each of these masses to the nearest whole number of kilograms.
    Write the approximate total mass for each set.

a   b   c

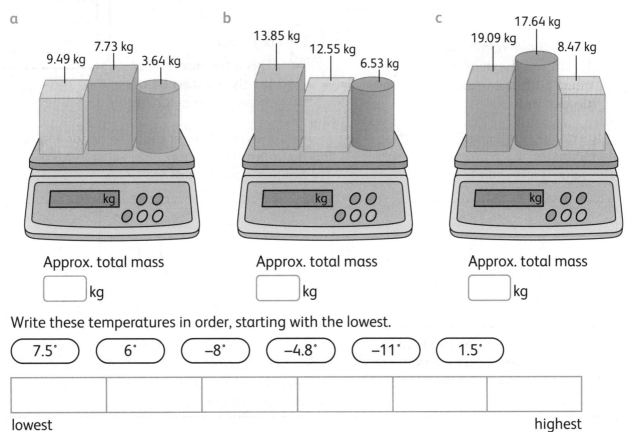

9.49 kg   7.73 kg   3.64 kg   13.85 kg   12.55 kg   6.53 kg   19.09 kg   17.64 kg   8.47 kg

Approx. total mass   Approx. total mass   Approx. total mass

☐ kg   ☐ kg   ☐ kg

4   Write these temperatures in order, starting with the lowest.

(7.5°)   (6°)   (−8°)   (−4.8°)   (−11°)   (1.5°)

lowest   highest

Rearrange this set of digits to make six different
decimal numbers between 1 and 10.
Use each digit only once in each decimal number.

☐ . ☐ ☐   ☐ . ☐ ☐

☐ . ☐ ☐   ☐ . ☐ ☐

☐ . ☐ ☐   ☐ . ☐ ☐

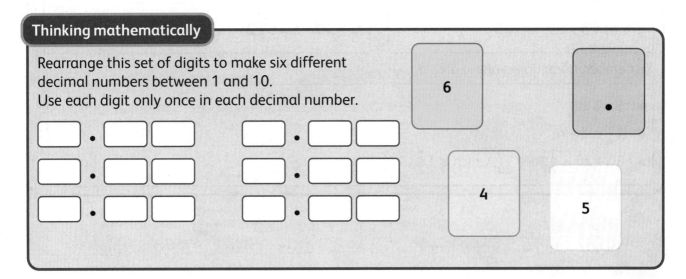

6

.

4

5

4.1   Fractions

Do you remember?

A **fraction** is a part of something that has been divided into equal parts.

This shape is divided into quarters. Three-quarters or $\frac{3}{4}$ is coloured green.

This is the **denominator**. It tells you the number of equal parts the whole is divided into.

$\frac{3}{4}$ ← numerator

denominator →

This is the **numerator**. It tells you the number of those parts you are using.

If the numerator is smaller than the denominator, it is a **proper fraction**. The value is less than 1.

If the numerator is larger than the denominator, it is an **improper fraction**. The value is greater than 1.

$\frac{1}{3}$

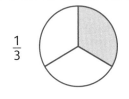

$\frac{3}{2}$

This is the same as the **mixed number** $1\frac{1}{2}$.

To change mixed numbers to improper fractions, multiply the denominator by the whole number and then add the numerator.

To change improper fractions to mixed numbers, divide the numerator by the denominator to find how many whole numbers there are and any fractions remaining.

$5\frac{3}{4} \rightarrow ?$

$(5 \times 4) + 3 = 23$ $5\frac{3}{4} = \frac{23}{4}$

$\frac{27}{5} \rightarrow ?$

$\frac{27}{5} = 27 \div 5 = 5 \text{ r } 2$ $\frac{27}{5} = 5\frac{2}{5}$

Remember, a fraction is also a division.

- $5 \div 8$ is $\frac{5}{8}$
- $1\frac{1}{4}$ is $5 \div 4$
- $30 \div 20$ is $\frac{30}{20}$ or $1\frac{10}{20}$, which is $1\frac{1}{2}$

Try this

1 What is $7 \div 10$ as a proper fraction?

    A $\frac{10}{7}$ B $\frac{7}{10}$ C $\frac{1}{3}$

2 What is $\frac{5}{3}$ as a mixed number?

    A $\frac{3}{5}$ B $2\frac{1}{3}$ C $1\frac{2}{3}$

3 What fraction is blue?

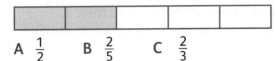

    A $\frac{1}{2}$ B $\frac{2}{5}$ C $\frac{2}{3}$

4 What is $3\frac{1}{2}$ as an improper fraction?

    A $\frac{5}{2}$ B $\frac{7}{2}$ C $\frac{6}{2}$

Practise

1 Write these as proper or improper fractions.

    a $7 \div 8 =$ ☐   b $12 \div 7 =$ ☐   c $90 \div 20 =$ ☐

    d $4 \div 15 =$ ☐   e $35 \div 9 =$ ☐   f $10 \div 18 =$ ☐

2 Write the fraction each arrow points to, as an improper fraction and a mixed number.

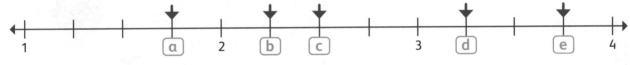

	a	b	c	d	e
Improper fraction					
Mixed number					

3 Change these to mixed numbers.

    a $\frac{13}{4} =$ ☐   b $\frac{28}{3} =$ ☐   c $\frac{45}{7} =$ ☐

    d $\frac{35}{6} =$ ☐   e $\frac{70}{9} =$ ☐   f $\frac{62}{12} =$ ☐

4 Change these to improper fractions.

    a $7\frac{2}{5} =$ ☐   b $19\frac{3}{4} =$ ☐   c $8\frac{2}{9} =$ ☐

    d $12\frac{7}{8} =$ ☐   e $15\frac{2}{3} =$ ☐   f $20\frac{1}{12} =$ ☐

Thinking mathematically

Use these numbers as a denominator or numerator:

   2   5   11   3   7

a Use pairs of the numbers to make fractions of less than $\frac{1}{2}$.

b Use pairs of these numbers to make improper fractions.

c Write your improper fractions as mixed numbers.

4.2 Fractions as operators

Do you remember?

When you need to find fractions of amounts, use the **numerator** and **denominator**.

What is $\frac{1}{5}$ of 40?
When the numerator is 1,
just divide by the denominator.

$\frac{1}{5}$ of 40 = 40 ÷ 5
$\qquad$ = 8

What is $\frac{3}{5}$ of 40?
When the numerator is more than 1,
divide by the denominator then multiply
by the numerator.

$\frac{1}{5}$ of 40 = 8

$\qquad \frac{3}{5} = \frac{1}{5} \times 3$

so, $\frac{3}{5}$ of 40 = 8 × 3 = 24

When an **improper fraction** is the **operator**, use the same method.
Divide by the denominator then multiply by the numerator.

What is $\frac{7}{5}$ of 40?

$\frac{1}{5}$ of 40 = 8

$\frac{7}{5} = \frac{1}{5} \times 7$

so, $\frac{7}{5}$ of 40 = 8 × 7 = 56

Use bar models
to help you.

If you are given the fraction of an amount and need to work out the total amount,
you can multiply to work it out. A bar model is helpful.

$\frac{3}{4}$ of a length of ribbon is 18 cm. What is the length of the whole ribbon?

$\frac{1}{4}$	$\frac{1}{4}$	$\frac{1}{4}$	$\frac{1}{4}$
6	6	6	

If we call the ribbon r:

$\frac{3}{4}$ of r = 18 $\qquad$ $\frac{1}{4}$ of r = 6

So, $\frac{4}{4}$ of r = 24

The whole length is 24 cm.

Maths words

numerator
denominator
improper fraction
operator

Try this

1 What is $\frac{1}{8}$ of 24?

  A 4 $\qquad$ B 6 $\qquad$ C 3

2 What is $\frac{2}{3}$ of 18?

  A 12 $\qquad$ B 3 $\qquad$ C 8

3 What is $\frac{3}{10}$ of 55?

  A 33 $\qquad$ B 16.5 $\qquad$ C 8.5

4 What is $\frac{5}{4}$ of 80?

  A 25 $\qquad$ B 90 $\qquad$ C 100

Practise

1   Answer each pair of questions.

a $\frac{1}{5}$ of 35 = ☐

   $\frac{2}{5}$ of 35 = ☐

b $\frac{1}{4}$ of 84 = ☐

   $\frac{3}{4}$ of 84 = ☐

c $\frac{1}{7}$ of 91 = ☐

   $\frac{5}{7}$ of 91 = ☐

d $\frac{1}{9}$ of 180 = ☐

   $\frac{8}{9}$ of 180 = ☐

e $\frac{1}{10}$ of 940 = ☐

   $\frac{12}{10}$ of 940 = ☐

f $\frac{1}{8}$ of 32 = ☐

   $\frac{9}{8}$ of 32 = ☐

2   Answer these.

a $\frac{1}{6}$ of n = 30

   n = ☐

b $\frac{1}{5}$ of x = 15

   x = ☐

c $\frac{3}{4}$ of a = 21

   a = ☐

d $\frac{2}{3}$ of y = 8

   y = ☐

e $\frac{3}{5}$ of n = 15

   n = ☐

f $\frac{5}{8}$ of x = 20

   x = ☐

3   Read and answer these.

a   Ali has \$64. He spends $\frac{3}{4}$ of his money on a new radio.
   How much does the radio cost?   ☐

b   What is $\frac{9}{10}$ of 80 kg?   ☐

c   An oil drum holds 60 litres, $\frac{2}{5}$ of the oil is poured into bottles.
   How much oil is poured into bottles?   ☐

d   A bag has 48 beads in it. $\frac{3}{8}$ of the beads are red. How many beads are red?   ☐

e   How many minutes are there in $\frac{3}{10}$ of one hour?   ☐

f   There are 25 oranges on a market stall. Emma bought $\frac{3}{5}$ of the oranges.
   How many did she buy?   ☐

Thinking mathematically

Sami takes $\frac{3}{4}$ of a box of 72 buttons.

Lisa takes $\frac{4}{5}$ of a box of 75 buttons.

a   Who has more buttons?

   ☐

b   How many buttons in total are left in the boxes?

   ☐ and ☐

4.3 Equivalent fractions

Do you remember?

Equivalent fractions have different **numerators** and **denominators** but are worth the same value.

$\frac{1}{3}$	$\frac{1}{3}$	$\frac{1}{3}$

=

$\frac{1}{6}$	$\frac{1}{6}$	$\frac{1}{6}$	$\frac{1}{6}$	$\frac{1}{6}$	$\frac{1}{6}$

Maths words
equivalent fractions
numerators
denominators
simplest form

A fraction can be changed into an equivalent fraction by multiplying the numerator and denominator by the same number.

$$\frac{3 \times 2 = 6}{5 \times 2 = 10} \qquad \frac{3 \times 4 = 12}{4 \times 4 = 16}$$

You can reduce a fraction to an equivalent fraction by dividing the top and bottom by the same number.

$$\frac{15 \div 5 = 3}{20 \div 5 = 4} \qquad \frac{40 \div 5 = 3}{50 \div 5 = 4}$$

$\frac{1}{12}$	$\frac{1}{12}$	$\frac{1}{12}$	$\frac{1}{12}$	$\frac{1}{12}$	$\frac{1}{12}$	$\frac{1}{12}$	$\frac{1}{12}$	$\frac{1}{12}$	$\frac{1}{12}$	$\frac{1}{12}$	$\frac{1}{12}$

$\frac{1}{6}$	$\frac{1}{6}$	$\frac{1}{6}$	$\frac{1}{6}$	$\frac{1}{6}$	$\frac{1}{6}$

$\frac{1}{2}$	$\frac{1}{2}$

$\frac{6}{12}, \frac{3}{6}$ and $\frac{1}{2}$ are equivalent fractions.

$\frac{1}{2}$ is the fraction in the **simplest form.**

Write $\frac{8}{12}$ in its simplest form.

$$\frac{8 \div 4 = 2}{12 \div 4 = 3}$$

Which of these is in its simplest form?

$$\frac{9}{15} \qquad \frac{3}{5} \qquad \frac{12}{20} \qquad \frac{6}{10}$$

Each fraction is equivalent to $\frac{3}{5}$, which is the simplest form.

Try this

1 What fraction is equivalent to $\frac{1}{3}$?

A $\frac{4}{9}$ B $\frac{3}{12}$ C $\frac{5}{15}$

2 What is the missing numerator? $\frac{3}{5} = \frac{?}{20}$

A 15   B 12   C 18

3 What fraction of this shape is green?

$\frac{1}{8}$	$\frac{1}{8}$	$\frac{1}{8}$	$\frac{1}{8}$	$\frac{1}{8}$	$\frac{1}{8}$	$\frac{1}{8}$	$\frac{1}{8}$

A $\frac{1}{4}$ B $\frac{3}{4}$ C $\frac{2}{3}$

4 What is $\frac{12}{15}$ in its simplest form?

A $\frac{3}{4}$ B $\frac{3}{5}$ C $\frac{4}{5}$

Practise

1 Complete these equivalent fractions.

a $\dfrac{2}{3} = \dfrac{\square}{12}$

b $\dfrac{4}{5} = \dfrac{12}{\square}$

c $\dfrac{3}{5} = \dfrac{\square}{45}$

d $\dfrac{3}{4} = \dfrac{18}{\square}$

2 Circle the odd one out in each set.

a $\dfrac{6}{24}$ $\dfrac{5}{20}$ $\dfrac{1}{5}$ $\dfrac{4}{16}$ $\dfrac{1}{4}$

b $\dfrac{4}{20}$ $\dfrac{3}{15}$ $\dfrac{5}{25}$ $\dfrac{4}{10}$ $\dfrac{1}{5}$

c $\dfrac{2}{3}$ $\dfrac{6}{9}$ $\dfrac{9}{12}$ $\dfrac{10}{15}$ $\dfrac{12}{18}$

d $\dfrac{9}{10}$ $\dfrac{4}{5}$ $\dfrac{16}{20}$ $\dfrac{8}{10}$ $\dfrac{12}{15}$

3 What fraction of each shape is shaded? Write each fraction in its simplest form.

a 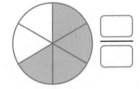 $\dfrac{\square}{\square}$

b 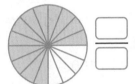 $\dfrac{\square}{\square}$

c 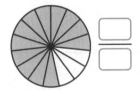 $\dfrac{\square}{\square}$

d  $\dfrac{\square}{\square}$

4 Complete these equivalent fraction chains.

a $\dfrac{2}{3} = \dfrac{4}{\square} = \dfrac{\square}{9} = \dfrac{\square}{\square}$

b $\dfrac{3}{4} = \dfrac{\square}{8} = \dfrac{9}{\square} = \dfrac{\square}{\square}$

c $\dfrac{4}{5} = \dfrac{8}{\square} = \dfrac{\square}{15} = \dfrac{\square}{\square}$

Thinking mathematically

The digits **2, 3, 4, 5, 6** and **8** are missing from these equivalent fractions.

Write each digit in its correct place.

$\dfrac{3}{\square} = \dfrac{\square}{10}$

$\dfrac{1}{\square} = \dfrac{\square}{8}$

$\dfrac{2}{\square} = \dfrac{\square}{12}$

Use each digit once only. Write the digits on small pieces of paper to help you.

37

4.4 Fractions, decimals and percentages

Do you remember?

Percentages are **fractions** out of 100.
'per cent' means 'out of 100' and % is the percentage sign.

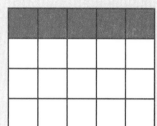

25% of the grid is red.

$\frac{5}{20} = \frac{25}{100} = 25\%$

25% is **equivalent to** 0.25 as a **decimal** number.

> **Maths words**
> percentages
> fractions
> equivalent to
> decimal

Look at these methods for converting between fractions, percentages and decimals:

1. Per cent to decimal

Divide the percentage by 100.

> 60% is equivalent to 0.6
> $60 \div 100 = 0.6$

2. Per cent to fraction

Write the percentage as a fraction out of 100 and then simplify.

> 40% is $\frac{40}{100}$, which is the same as $\frac{2}{5}$

3. Decimal to per cent

Multiply the decimal by 100.

> 0.25 is equivalent to 25%
> $0.25 \times 100 = 25$

4. Fraction to per cent

Write the fraction as a decimal and then multiply by 100.

> $\frac{3}{4}$ is 0.75, which is the same as 75%
> $0.75 \times 100 = 75$

5. Decimal to fraction

Write it as a hundredth and then in its simplest form.

> 0.6 is equivalent to $\frac{3}{5}$
> $0.6 = \frac{60}{100} = \frac{6}{10} = \frac{3}{5}$

6. Fraction to decimal

Divide the numerator by the denominator.

> $\frac{7}{8} = 0.875$
> $$\begin{array}{r} 0\,.\,8\;7\;5 \\ 8\,\overline{)7\,.\,0\;0\;0} \end{array}$$

Try this

1   What is $\frac{13}{25}$ as a percentage?

    A 13%   B 42%   C 52%

2   What is 40% as a fraction in its simplest form?

    A $\frac{1}{4}$ B $\frac{3}{10}$ C $\frac{2}{5}$

3   What is 0.08 as a percentage?

    A 80%   B 8%   C 0.08%

4   What is 0.45 as a fraction in its simplest form?

    A $\frac{4}{5}$ B $\frac{3}{8}$ C $\frac{9}{20}$

Practise

1 Write the missing digits to complete these.

a $\dfrac{1}{\boxed{}} = 0.\boxed{} = 50\%$

b $\dfrac{\boxed{}}{4} = 0.25 = \boxed{}\%$

c $\dfrac{1}{20} = 0.\boxed{} = \boxed{}\%$

d $\dfrac{2}{\boxed{}} = 0.4 = \boxed{}\%$

e $\dfrac{17}{50} = 0.\boxed{} = \boxed{}\%$

f $\dfrac{7}{\boxed{}} = 0.7 = \boxed{}\%$

g $\dfrac{\boxed{}}{100} = 0.09 = \boxed{}\%$

h $\dfrac{11}{\boxed{}} = 0.44 = \boxed{}\%$

2 Look at this diagram.

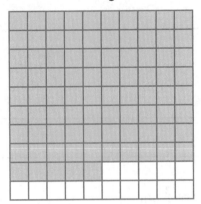

a What fraction, in its simplest form, is shaded? $\boxed{}$

b Write this as a decimal. $\boxed{}$

c Write this as a percentage. $\boxed{}$

4 Joseph wants to compare his scores in these 5 maths tests. Change them all to percentages to work out which test he scored highest and which was his lowest score.

Test	Score	Percentage
1	$\dfrac{7}{10}$	
2	$\dfrac{18}{20}$	
3	$\dfrac{4}{5}$	
4	$\dfrac{21}{25}$	
5	$\dfrac{38}{50}$	

3 Write <, > or = to make each statement true.

a 25% $\boxed{}$ 0.25

b 0.2 $\boxed{}$ $\dfrac{1}{2}$

c 4% $\boxed{}$ 0.4

d 0.1 $\boxed{}$ 10%

e 7% $\boxed{}$ $\dfrac{7}{10}$

f $\dfrac{2}{5}$ $\boxed{}$ 0.04

Thinking mathematically

A school sends 200 children on a school trip, 80 girls and 120 boys.

a What fraction are boys? $\boxed{}$

b What percentage are boys? $\boxed{}$

c What fraction are girls? $\boxed{}$

d What percentage are girls? $\boxed{}$

e In the group of children, $\dfrac{1}{5}$ are under 8 years old. What percentage of the group is 8 years or older? $\boxed{}$

4.5 Addition and subtraction of fractions

Maths words

denominators

unlike fractions

like fractions

equivalent fractions

common denominator

simplest form

Do you remember?

Fractions with different **denominators** are **unlike fractions**.

To add or subtract unlike fractions, we change them to **like fractions** by looking for **equivalent fractions** with a **common denominator**.

Follow these steps:

1   Find equivalent fractions with a common, or same, denominator.

2   Add or subtract the numerators and write the numerator over the common denominator.

3   Write the fraction in its **simplest form** if needed.

Example

What is $\frac{1}{6} + \frac{2}{3}$?

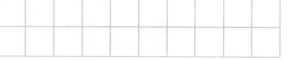

6 is a multiple of 3, so 6 is a common denominator of $\frac{1}{6}$ and $\frac{2}{3}$.

$\frac{1}{6} + \frac{4}{6} = \frac{5}{6}$ So, $\frac{1}{6} + \frac{2}{3} = \frac{5}{6}$

Example

What is $\frac{3}{5} - \frac{1}{4}$?

20 is a multiple of both 4 and 5, so 20 is a common denominator of $\frac{3}{5}$ and $\frac{1}{4}$.

$\frac{3}{5} = \frac{12}{20}$ $\frac{1}{4} = \frac{5}{20}$

$\frac{12}{20} - \frac{5}{20} = \frac{7}{20}$ So, $\frac{3}{5} - \frac{1}{4} = \frac{7}{20}$

Try this

1   What is $\frac{1}{3} + \frac{1}{4}$?

   A $\frac{1}{7}$ B $\frac{1}{12}$ C $\frac{7}{12}$

2   What is $\frac{1}{2}$ added to $\frac{1}{10}$?

   A $\frac{3}{5}$ B $\frac{1}{5}$ C $\frac{7}{10}$

3   What is $\frac{1}{3}$ subtract $\frac{1}{6}$?

   A $\frac{2}{9}$ B $\frac{1}{2}$ C $\frac{1}{6}$

4   What is $\frac{1}{2} - \frac{2}{5}$?

   A $\frac{1}{10}$ B $\frac{1}{3}$ C $\frac{1}{5}$

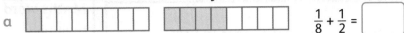

Practise

1   Rewrite the fractions so that they have the same denominator, then add the fractions.

   a $\frac{1}{8} + \frac{1}{2} =$ ⬚

   b $\frac{1}{3} + \frac{2}{5} =$ ⬚

   c $\frac{1}{2} + \frac{3}{8} =$ ⬚

   d $\frac{3}{4} + \frac{1}{3} =$ ⬚

2   Rewrite the fractions so that they have the same denominator, then subtract the fractions.

   a $\frac{3}{4} - \frac{1}{8} =$ ⬚

   b $\frac{3}{10} - \frac{1}{5} =$ ⬚

   c $\frac{2}{3} - \frac{1}{2} =$ ⬚

   d $\frac{5}{6} - \frac{3}{4} =$ ⬚

3   Find the common denominator for each pair of fractions.
   Write the answers in their simplest form.

   a $\frac{1}{8} + \frac{1}{4} =$ ⬚

   b $\frac{4}{5} - \frac{1}{3} =$ ⬚

   c $\frac{7}{10} + \frac{1}{3} =$ ⬚

   d $\frac{7}{4} - \frac{5}{8} =$ ⬚

   e $\frac{3}{8} + \frac{3}{2} =$ ⬚

   f $\frac{6}{5} - \frac{3}{4} =$ ⬚

4   Read and answer these.

   a   A farmer buys some land. He plants $\frac{1}{3}$ with plantain and $\frac{1}{5}$ with maize.
       What fraction of his land has he planted so far?

   b   After a meal, $\frac{2}{5}$ of a vegetable pizza and $\frac{3}{10}$ of a spicy meat pizza is left.
       What fraction of pizza is left altogether?

   c   A floor is made of $\frac{1}{3}$ red tiles and $\frac{1}{4}$ yellow tiles. The rest of the tiles are blue.
       What fraction of the tiles are blue?

   d   A jug is filled $\frac{1}{2}$ full with orange juice and $\frac{1}{3}$ with pineapple juice.
       What fraction of the jug is left to be filled?

⬚
⬚
⬚
⬚

Thinking mathematically

Copy and complete these. Use each of the digits 1 to 6 to fill in the boxes.

1   2   3   4   5   6

$\frac{⬚}{10} + \frac{⬚}{5} = \frac{7}{10}$

$\frac{1}{⬚} + \frac{1}{3} = \frac{⬚}{2}$

$\frac{3}{8} + \frac{1}{⬚} = \frac{⬚}{8}$

41

4.6 Multiplication and division of fractions

Do you remember?

When we **multiply** and divide fractions, we can use diagrams to help us solve the problems.

Rosa has three bottles, each with $\frac{3}{4}$ litre of orange juice.

She pours it all together in a jug. How many litres of orange juice are in the jug?

$\frac{3}{4} \times 3 = \boxed{}$

This diagram shows the three bottles, with $\frac{3}{4}$ shaded for each:

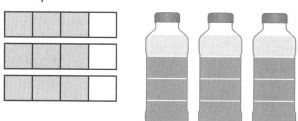

There are 9 quarters, so these could be put together:

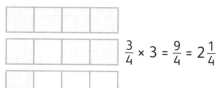

$\frac{3}{4} \times 3 = \frac{9}{4} = 2\frac{1}{4}$

Do you notice that you can multiply the numerator by the multiplier to find the answer? Check to see if this always work.

When we divide fractions, try to use a diagram to help make sense of the problem.

Mohan shares $\frac{3}{5}$ of a bar of chocolate between a group of four.
What fraction of the bar of chocolate do they each get?

$\frac{3}{5} \div 4 = \boxed{}$

Divide a bar into fifths and shade $\frac{3}{5}$:

$\frac{1}{5}$	$\frac{1}{5}$	$\frac{1}{5}$	$\frac{1}{5}$	$\frac{1}{5}$

Divide the bars into four to show the fraction each person gets:

$\frac{1}{20}$	$\frac{1}{20}$	$\frac{1}{20}$		

Each person gets $\frac{3}{20}$ of the bar of chocolate.

$\frac{3}{5} \div 4 = \frac{3}{20}$

Try this

1 What is $\frac{1}{3} \times 5$?

    A $\frac{1}{15}$ B $1\frac{2}{3}$ C $\frac{3}{5}$

2 What is $\frac{1}{2} \div 3$?

    A $\frac{2}{3}$ B 6   C $\frac{1}{6}$

3 What is $\frac{3}{4}$ multiplied by 6?

    A 5   B 9   C $4\frac{1}{2}$

4 What is $\frac{2}{5}$ divided by 2?

    A $\frac{1}{5}$ B $\frac{4}{5}$ C $\frac{1}{2}$

Practise

1 Write a multiplication sentence to match each diagram.

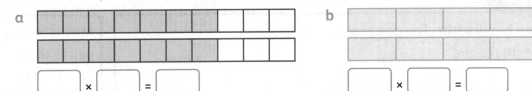

a $\boxed{} \times \boxed{} = \boxed{}$

b $\boxed{} \times \boxed{} = \boxed{}$

2 Answer these.

a $\frac{1}{3} \times 6 = \boxed{}$

b $\frac{1}{10} \times 5 = \boxed{}$

c $\frac{3}{4} \times 5 = \boxed{}$

d $\frac{4}{5} \times 8 = \boxed{}$

e $\frac{5}{8} \times 3 = \boxed{}$

f $\frac{7}{10} \times 5 = \boxed{}$

3 Use the diagrams to help you answer these.

a $\frac{1}{4} \div 3 = \boxed{}$

b $\frac{2}{3} \div 2 = \boxed{}$

c $\frac{2}{5} \div 4 = \boxed{}$

d $\frac{3}{8} \div 2 = \boxed{}$

4 Answer these.

a $\frac{1}{2} \div 3 = \boxed{}$

b $\frac{1}{3} \div 5 = \boxed{}$

c $\frac{3}{4} \div 4 = \boxed{}$

d $\frac{3}{5} \div 3 = \boxed{}$

e $\frac{2}{3} \div 6 = \boxed{}$

f $\frac{3}{10} \div 2 = \boxed{}$

5 Solve these problems.

a The length of one side of a square field is $\frac{2}{5}$ km.

What is the length of the perimeter of the field?

b Tanya has a ribbon that is $\frac{3}{5}$ m long. She cuts it into 6 equal lengths.

What is the length of each of these pieces?

c Femi uses $\frac{5}{8}$ kg of mangoes to make one fruit smoothie.

What mass of mangoes does he need to make six fruit smoothies?

Thinking mathematically

These are the dimensions of three rectangles. Estimate and then calculate the area of each.

Rectangle	Length	Width	Estimate of area	Calculation	Area
A	3 m	$\frac{7}{10}$ m	____m²		____m²
B	5 m	$\frac{3}{5}$ m	____m²		____m²
C	6 m	$\frac{7}{8}$ m	____m²		____m²

4.7 Percentages

Do you remember?

Percentages are **hundredths**. The percentage sign is %.

Common percentages are:

$10\% = \frac{10}{100} = \frac{1}{10}$ $20\% = \frac{20}{100} = \frac{1}{5}$

$25\% = \frac{25}{100} = \frac{1}{4}$ $75\% = \frac{75}{100} = \frac{3}{4}$

Maths words

percentages
hundredths

We often need to work out percentages of amounts.

What is 20% of $60?

Method 1	**Method 2**
Change to a fraction and multiply: $20\% = \frac{1}{5}$ $\frac{1}{5} \times 60 = \frac{60}{5} = \$12$	Use 10% to work it out. Just divide by 10: 10% of $60 is $6. So, 20% of $60 is double that: $12

If you know how to work out percentages of amounts, you can work out sale prices. There are two steps to remember.

A shop has a 25% sale. The original price of a flag is $30. What is the sale price?

Step 1: Work out the percentage: 25% of $30 is $7.50.

Step 2: Subtract this amount from the price: $30 − $7.50 = $22.50

The sale price of the flag is $22.50.

If it is an increase, there are still two steps, but you add the percentage to the amount.

3 litre bottles of oil are increased in size by 20%. What is the new capacity of the bottles?

Step 1: Work out the percentage: 20% of 3 litres is 0.6 litres.

Step 2: Add this amount to the original: 3ℓ + 0.6ℓ = 3.6 litres

The bottles now hold 3.6 litres of oil.

Try this

1   What is 40% of $200?

    A $80   B $160   C $60

2   A jug holds 800 ml. If it is 75% full of water, how much water is in the jug?

    A 200 ml   B 750 ml   C 600 ml

3   A $32 shirt is reduced in price by 5%. What is the new price of the shirt?

    A $16   B $30.40   C $28.60

4   In a class of 45 children, 60% are boys. How many girls are there?

    A 18   B 27   C 21

Practise

1 Write the percentages of each of these amounts.

   a $50   b $80   c $25   d $200
      10% → $ [   ]   10% → $ [   ]   10% → $ [   ]   10% → $ [   ]
      30% → $ [   ]   40% → $ [   ]   30% → $ [   ]   40% → $ [   ]

2 Write these amounts.

   a 10% of 70 cm is _____   b 30% of 90 km is _____
   c 20% of 20 litres is _____   d 40% of 30 kg is _____
   e 25% of 80 m is _____   f 50% of 400 mm is _____

3 Calculate the reduced price of each item.

   a $48 SALE 25% off   b $23 SALE 50% off   c $25.50 SALE 10% off
      $ [   ] $ [   ] $ [   ]

   d $20 SALE 40% off   e $12.50 SALE 20% off   f $5.20 SALE 5% off
      $ [   ] $ [   ] $ [   ]

4 Answer these questions.

   a The price of a bike is increased by 15% from $220.
      What is the new price?   [   ]

   b A bike helmet costs $32. The price increases by 30%.
      What is the new price?   [   ]

   c A set of bicycle lights has a price rise of 25%. The original cost was $48.
      What is the new price of the lights?   [   ]

Thinking mathematically

At the end of a week, a shopkeeper counted up all the money in the till.
There were exactly 200 notes or coins in total.

25% are **$1**   30% are **$5**   20% are **$10**
5% are **$20**   10% are **$100**   The rest are **50 cent** coins

Write how many of each coin or note there is in the till.

$1 → [   ] $5   → [   ] $10 → [   ]
$20 → [   ] $100 → [   ]   50c → [   ]

How much money is there altogether?
[   ]

4.8 Comparing and ordering fractions

Do you remember?

The symbols > and < are used to **compare** numbers, including fractions.

< means 'is less than' $\frac{1}{3} < \frac{2}{3}$

> means 'is greater than' $\frac{1}{2} > \frac{1}{10}$

If fractions have the same **denominator**, they are easy to compare.

$\frac{3}{5}$ > $\frac{2}{5}$

To compare any fractions change them to **equivalent fractions** with a **common denominator**. This means they have the same denominator.

Which is the greater fraction, $\frac{2}{3}$ or $\frac{3}{4}$?

Find the equivalent fractions to $\frac{2}{3}$ and $\frac{3}{4}$ that have a common denominator:

$\frac{2}{3} = \frac{4}{6} = \frac{6}{9} = \frac{8}{12}$

$\frac{3}{4} = \frac{6}{8} = \frac{9}{12}$

$\frac{9}{12}$ is greater than $\frac{8}{12}$

$\frac{3}{4}$ is greater than $\frac{2}{3}$

$\frac{3}{4} > \frac{2}{3}$

Use the same method to put a group of fractions in order of size.

Put these fractions in order, starting with the smallest: $\frac{1}{2}, \frac{1}{4}, \frac{3}{8}$.

Find equivalent fractions with a common denominator:

$\frac{1}{2} = \frac{4}{8}$ $\frac{1}{4} = \frac{2}{8}$ $\frac{3}{8} = \frac{3}{8}$

Put them in order:

$\frac{1}{4}$ $\frac{3}{8}$ $\frac{1}{2}$

Maths words

compare
denominator
equivalent fractions
common denominator

Try this

1 Which of these fractions is greater than $\frac{3}{4}$?
   $\boxed{} > \frac{3}{4}$
   A $\frac{7}{10}$ B $\frac{7}{8}$ C $\frac{3}{5}$

2 Which of these fractions is greater than $\frac{2}{5}$?
   $\boxed{} > \frac{2}{5}$
   A $\frac{3}{10}$ B $\frac{3}{8}$ C $\frac{3}{4}$

3 Which of these fractions is smaller than $\frac{3}{10}$?
   $\boxed{} < \frac{3}{10}$
   A $\frac{2}{5}$ B $\frac{2}{3}$ C $\frac{1}{4}$

4 Which of these fractions is smaller than $\frac{5}{8}$?
   $\boxed{} < \frac{5}{8}$
   A $\frac{3}{10}$ B $\frac{7}{10}$ C $\frac{4}{5}$

Practise

1   Write < or > between each pair of fractions.

a $\frac{2}{3}$ ☐ $\frac{1}{2}$ b $\frac{1}{4}$ ☐ $\frac{1}{3}$ c $\frac{2}{3}$ ☐ $\frac{4}{5}$

d $\frac{1}{2}$ ☐ $\frac{2}{5}$ e $\frac{3}{4}$ ☐ $\frac{5}{6}$ f $\frac{3}{10}$ ☐ $\frac{1}{5}$

2   Put each group of fractions in order starting with the smallest.

a $\frac{1}{4}$ $\frac{1}{2}$ $\frac{3}{8}$ $\frac{3}{4}$ ☐

b $\frac{1}{6}$ $\frac{3}{4}$ $\frac{1}{3}$ $\frac{5}{12}$ ☐

c $\frac{2}{3}$ $\frac{1}{2}$ $\frac{3}{5}$ $\frac{7}{10}$ ☐

d $\frac{9}{10}$ $\frac{4}{5}$ $\frac{1}{2}$ $\frac{1}{4}$ ☐

3   Write the digits 3, 4, 5, 6, 7 and 8 in the boxes. Each digit can only be used once.
    Each fraction must be a proper fraction in its simplest form.
    Write the numbers on small pieces of paper to help you.

$\frac{4}{\Box}$ > $\frac{2}{\Box}$ > $\frac{3}{\Box}$ $\frac{1}{\Box}$ < $\frac{3}{\Box}$ < $\frac{6}{\Box}$

Thinking mathematically

Compare the fractions represented by each colour in the picture.

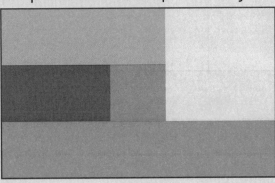

a   Write a fraction for each colour in its simplest form.

Red → ☐   Yellow → ☐   Orange → ☐   Green → ☐   Blue → ☐

b   Write the fractions in order starting with the smallest.

☐ < ☐ < ☐ < ☐ < ☐

4.9 Multiplication and division of decimals

Do you remember?

There are two ways to work out the **product** when multiplying **decimals**. Remember to always **estimate** the answer first.

Method 1:

Partition the decimal fraction into the whole number and the decimal. Multiply each part separately, then add the two numbers together.

Method 2:

The number of decimal places in the product is the same as the number of decimal places in the number multiplied.
So, multiply as with whole numbers, but check that the decimal point and the digits give the correct **place value**.

37.48 × 9	Method 1	Method 2
Estimate: $\approx 37 \times 10$ ≈ 370	$= (37 \times 9) + (0.48 \times 9)$ $= 333 + 4.32$ $= 337.32$	$\begin{array}{r} 37.48 \\ \times \quad 9 \\ \hline 337.32 \end{array}$

Maths words
product
decimals
estimate
place value
dividend

When you divide decimals by whole numbers, use the same method as dividing with whole numbers. Always check the place value of the digits. With short division, put a decimal point in the answer directly above the decimal point of the **dividend**.
Estimate the answer first so that you can check if it is close.

| $18.25 \div 5 = \boxed{}$ **Estimate:** $\approx 20 \div 5$ ≈ 4 | $\begin{array}{r} 3.65 \\ 5\,)\overline{18.25} \\ -15 \\ \hline 3.25 \\ -\ 3.00 \\ \hline 0.25 \end{array}$ $18.25 \div 5 = 3.65$ | $\begin{array}{r} 3.65 \\ 5\,)\overline{18.25} \end{array}$ |

Try this

1 2.45 × 8 = ?

   A 16.45   B 19.6   C 18.32

2 What is 32.8 multiplied by 6?

   A 196.8   B 186.4   C 193.8

3 What is 16.88 divided by 4?

   A 4.22   B 4.11   C 4.02

4 What is the missing number?

   $3.24 \div \boxed{} = 1.08$

   A 2   B 30   C 3

Practise

1 Estimate then complete these.

a	b	c	d
4.95	16.48	194.9	94.83
× 5	× 3	× 7	× 8

2 Estimate first, then answer these.

a 3.85 × 8 = ☐   b 4.09 × 7 = ☐

c 15.97 × 3 = ☐   d 126.3 × 9 = ☐

3 Estimate first, then complete these.

a 9)‾272.7‾ = ☐   b 5)‾38.15‾ = ☐   c 6)‾73.2‾ = ☐   d 8)‾198.4‾ = ☐

4 Estimate first, then answer these.

a 64.2 ÷ 6 = ☐   b 58.87 ÷ 7 = ☐   c 707.4 ÷ 3 = ☐   d 323.91 ÷ 9 = ☐

5 Answer these. Remember to estimate the answer first.

a A metre of rope weighs 2.67 kg. How much does 8 m weigh? ☐

d A dining chair costs $157.80. What is the total cost of 6 chairs? ☐

c A path is 16.32 m long and is made with 8 square paving slabs. What is the length of each paving slab? ☐

d A builder buys 8 planks of wood. They weigh 144.64 kg in total. How much does one plank of wood weigh? ☐

Thinking mathematically

Answer this problem.
A group of 9 people book a trip to a museum.
It costs $6.45 for adults and $3.50 for children.
The group pays a total of $43.30.
How many children are in the group?

Number	Adults	Children	Total
1	$6.45	$3.50	
2			
3			
4			
5			
6			

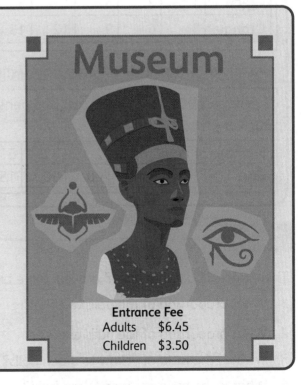

Museum

Entrance Fee
Adults $6.45
Children $3.50

4.10 Ratio and proportion

Maths words
proportion
ratio
direct proportion

Do you remember?

When you look at the **proportion** of an amount, it is the same as finding the fraction of the whole amount.

What proportion of the tiles are white?

There are 8 tiles altogether, 2 of them are white, so $\frac{2}{8}$ of the tiles are white. $\frac{2}{8}$ can be simplified to $\frac{1}{4}$. So, the proportion of white tiles is 1 in every 4, or $\frac{1}{4}$.

This can be used to work out the number of white tiles for a larger tile pattern, because the proportion of $\frac{1}{4}$ stays the same.

Total number of tiles	8	16	24	32	40
White tiles	2	4	6	8	10

Ratio is a little different to proportion because it compares one amount with another. It shows the relationship between two or more amounts.

What is the ratio of green to orange tiles?

There are 3 green tiles and 9 orange tiles.

The ratio is 3 to 9 or $\frac{3}{9}$. This can be written as $3:9$.

We can simplify ratios. $\frac{3}{9} = \frac{1}{3}$ (both terms divide by 3).

So, the ratio of green to orange is 1 to 3, or $1:3$.
For every 1 green tile, there are 3 orange tiles.
This ratio stays the same for different amounts:

How many green tiles will be needed to tile a wall with 40 tiles?

Green	1	2	3	4	5
Orange	3	6	9	12	15

Two quantities are in **direct proportion** when they both increase or decrease in the same ratio.

3 pens cost 90c. What is the cost of 5 pens?

30c × 5 = $1.50

Work out the cost of 1 pen.

Number of Pens	1	2	3	5
Cost		30c	60c	90c	150c

Try this

There are 20 balls in a bag. 8 are red, 5 are blue, 6 are green and 1 is yellow.

1 What proportion of the balls are blue?   A $\frac{1}{5}$ B $\frac{1}{3}$ C $\frac{1}{4}$

2 What proportion of the balls are red?   A $\frac{2}{5}$ B $\frac{1}{8}$ C $\frac{2}{3}$

3 What is the ratio of yellow to green balls?   A $1:20$ B $1:6$ C $1:19$

4 What is the ratio of green to red balls?   A $3:7$ B $4:3$ C $3:4$

Practise

1   Look at these tile patterns. What proportion of each of the patterns is blue?

a   b   c   d   e   f

$\frac{\square}{\square}$ $\frac{\square}{\square}$ $\frac{\square}{\square}$ $\frac{\square}{\square}$ $\frac{\square}{\square}$ $\frac{\square}{\square}$

2   What is the ratio of blue to orange tiles in each of the patterns above?

a $\square : \square$ b $\square : \square$ c $\square : \square$

d $\square : \square$ e $\square : \square$ f $\square : \square$

3   Read and answer these.

a   In a class there are 2 girls to every 3 boys.
    There are 30 children in the class.
    How many boys are there?

b   The ratio of red balloons to green balloons in a pack is 2:5.
    One pack has 12 red balloons.
    How many green balloons are in the pack?

c   The proportions of a door are $\frac{1}{3}$ width to height.
    If the door is 270 cm high, what is the width?

d   A bucket holds 12 litres of water. A glass holds 600 ml of water.
    What is the ratio of the amount a bucket holds to the
    amount a glass holds?

4   Use direct proportion to answer these.

a   Six toy cars cost a total of $4.50. What is the cost of ten cars?

b   Four bars of soap cost a total of $3.20.
    What is the cost of three bars of soap?

c   A typist types 400 words in 8 minutes.
    How many words does she type in 5 minutes?

d   Five bags of sugar weigh 4 kg.
    What is the total mass of six bags of sugar?

Thinking mathematically

Here is part of a recipe for four people. Write the recipe for six people.

180 g lentils $\square$ 2 tablespoons melted butter $\square$

120 g split peas $\square$ 1 medium onion $\square$

580 ml water $\square$ $\frac{3}{4}$ spoonful mixed herbs $\square$

Test your understanding

1 The rule for a number sequence is 'multiply by 2 and subtract 3'. What is the next number in the sequence?

1 mark

4	5	7	11	

2 Look at these two numbers.

2 marks

8.07	14.8

a What is the difference between them? ☐

b What is the total? ☐

3 Write these numbers in the correct place on this Carroll diagram.

2 marks

18	30	12	24

	Multiple of 4	Not a multiple of 4
Factor of 60		
Not a factor of 60		

4 What is 0.2 × 0.2 × 0.2? Circle the correct answer.

1 mark

A 0.06   B 0.0008   C 0.008   D 0.8

5 Write the value of each letter.

2 marks

a $7n - 5 = 51$ $n = $ ☐

b $10 + (36 \div t) = 14$ $t = $ ☐

6 Three friends bought some food from a cafe:

RECEIPT 1
****
1 cake
1 roll

Total: $1.85
****

RECEIPT 2
****
2 cakes
1 sandwich

Total: $2.95
****

RECEIPT 3
****
1 sandwich
2 rolls

Total: $3.65
****

What is the cost of each item?

a  cake = ☐   sandwich = $1.45   b roll = ☐

2 marks

Test your understanding

7 Write the missing numbers.   **2 marks**

   a   20% of 80 is ☐

   b   20% of ☐ is 80

8 List the common factors for each of these.   **2 marks**

   a   Common factors of 45 and 30   →   _____

   b   Common factors of 18 and 54   →   _____

9 Write the missing numbers.   **2 marks**

   a   (☐ × 2) ÷ 3 = 8

   b   (☐ × 4) − (54 ÷ 6) = 7

10 Write these fractions in order, starting with the smallest: **1 mark**

$\frac{3}{4}$	$\frac{1}{3}$	$\frac{1}{4}$	$\frac{5}{12}$	$\frac{2}{3}$
☐	☐	☐	☐	☐

11 Complete these: **2 marks**

   a
```
      ☐  9  .  5  3
   +  1  5  .  1  ☐
   _____
      4  ☐  .  6  7
```

   b
```
      5  ☐  .  0  9
   −  2  7  .  ☐  6
   _____
      2  5  .  1  ☐
```

12 Circle the division that has a remainder of 1.   **1 mark**

   A   281 ÷ 6   B   347 ÷ 3   C   237 ÷ 4   D   695 ÷ 8

13 In a carnival procession, there are 3 girls and 2 boys in each line.   **1 mark**
       There are 42 girls taking part. How many boys are in the procession?   ☐

14 A class is collecting money for charity. They want a total of $1000.   **2 marks**
       By the end of May they have collected $800.
       What percentage have they collected by the end of May?   ☐

15 Write the missing numbers.   **2 marks**

   a $\frac{3}{4}$ of 100 = $\frac{1}{2}$ of ☐

   b $\frac{2}{3}$ of 60 = $\frac{1}{4}$ of ☐

Total: _____ marks out of 25

53

5.1 Comparing times

Do you remember?

Timetables and **digital** watches often use the **24-hour clock**.

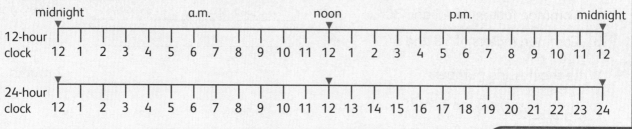

| | midnight | | | | | a.m. | | | | | | noon | | | | | p.m. | | | | | midnight |

12-hour clock 12 1 2 3 4 5 6 7 8 9 10 11 12 1 2 3 4 5 6 7 8 9 10 11 12

24-hour clock 12 1 2 3 4 5 6 7 8 9 10 11 12 13 14 15 16 17 18 19 20 21 22 23 24

6.35 a.m. → 06:35	10.49 a.m. → 10:49
6.35 p.m. → 18:35	10.49 p.m. → 22:49

Maths words

timetable

digital

24-hour clock

a.m.

p.m.

a.m. is 'ante meridiem' and means 'before midday'.
p.m. is 'post meridiem' and means 'after midday'.

The world is divided into 24 time zones, each set one hour apart.
Some countries have more than one time zone.

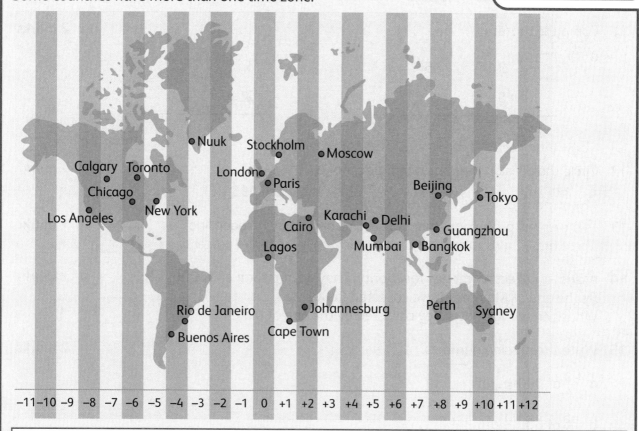

−11 −10 −9 −8 −7 −6 −5 −4 −3 −2 −1 0 +1 +2 +3 +4 +5 +6 +7 +8 +9 +10 +11 +12

The time difference between New York and Moscow is 8 hours.
When it is midday in New York, it is 8 p.m. in Moscow. Moscow is 8 hours ahead of New York.
When it is midday in Moscow, it is 4 a.m. in New York. New York is 8 hours behind Moscow.

Try this

1 What time is 11:38 p.m. using 24-hour time?

    A 11:38   **B** 21:38   **C** 23:38

2 What time is it in Cairo if it is 08:40 in London?

    A 06:40   **B** 08:40   **C** 10:40

3 What time is 02:51 using 12-hour time?

    A 4.51 p.m.   **B** 2.51 a.m.   **C** 2.51 p.m.

4 What is the time difference between Paris and Tokyo?

    A 8 hours   **B** 9 hours   **C** 10 hours

Practise

1 Write these times as 24-hour clock times.

  a 10.15 a.m.   ☐:☐   b 08.55 a.m.   ☐:☐

  c 04.00 p.m.   ☐:☐   d 07.30 p.m.   ☐:☐

2 Write these times as 12-hour clock times, using a.m. and p.m.

  a 14:50   ☐.☐☐   b 11:08   ☐.☐☐

  c 07:22   ☐.☐☐   d 23:10   ☐.☐☐

3 Answer these questions.

  a How many days are there in sixteen weeks? ☐

  b How many seconds are there in 5.5 minutes? ☐

  c How many hours are there in 2.5 days? ☐

  d How many days are there in March? ☐

  e How many seconds are there in 4.5 hours? ☐

4 Read these sentences and explain why they each made a mistake.

  a Jose checks a timetable. His train leaves at 14:50. He says, 'My train leaves at 4.50 p.m.'

  b Ben checks the time. It is 23:45. He says, 'It will be 24:15 in half an hour.'

Thinking mathematically

Use the world time chart to identify which statements are true and which are false.
Explain your reasoning for each statement.

a When it is 6.00 p.m. in Karachi, it is 2.00 p.m. in London.   True/False

b When it is 03:00 in Chicago, it is 10:00 in Cape Town.   True/False

c 08:00 is the time in Perth when 05:00 is the time in Cairo.   True/False

d When it is 9 a.m. in Sydney on 24 March, in London it is still 23 March.   True/False

5.2 Time intervals

Do you remember?

Drawing a timeline and counting on it can help to calculate time **intervals**.

Maths words
intervals
decimal
equivalent to

A bus leaves at 13:52 and arrives at 15:37. How long is the journey?

13:52 14:00 15:00 **15:37**

8 mins 1 hour 37 mins

Total time for the journey is 1 hour + 8 minutes + 37 minutes = 1 hour 45 minutes

Time can be written in hours and minutes or as a **decimal**.
1 hour 45 minutes is **equivalent to** 1.75 hours because 0.75 × 60 = 45 minutes.
Here are the key equivalent times:

0.1 h	0.2 h	0.25 h	0.3 h	0.4 h	0.5 h	0.6 h	0.7 h	0.75 h	0.8 h	0.9 h
6 min	12 min	15 min	18 min	24 min	30 min	36 min	42 min	45 min	48 min	54 min

Try this

1 A TV programme starts at 15:52 and ends 35 minutes later. What time does it finish?
   A 16:27   B 16:43   C 16:33

2 A TV programme starts at 17:45 and ends at 18:32. How long is the programme?
   A 1 hour 13 minutes   B 47 minutes   C 1 hour 43 minutes

3 4.3 hours is the same as:
   A 4 hours 30 mins   B 4 hours 3 mins   C 4 hours 18 mins

4 1 hr 15 mins is the same as:
   A 1.15 hours   B 1.3 hours   C 1.25 hours

Practise

1 Use the timeline and write how long between each time.

a **09:35**   10:00   11:00   **11:12**   total time: [   ]

b **20:49**   21:00   22:00   23:00   **23:26**   total time: [   ]

c **22:51**   23:00   00:00   01:00   **01:37**   total time: [   ]

2   Write the length of time between these pairs of times.

a   11:58 → 12:38 ☐   b   17:45 → 19:07 ☐

c   13:52 → 15:03 ☐   d   23:50 → 00:26 ☐

3   Read and answer these time-interval problems.

a   Amir watched TV for 2.5 hours until 21:05.
    What time did he start watching TV?   ☐

b   A ferry timetable shows that the boat leaves at 07.30 for a 4.2 hr journey.
    What time does the boat arrive?   ☐

c   A film starts at 6.45 p.m. and runs for 102 minutes.
    What times does the film end?   ☐

d   A cake needs to cook for 55 minutes. It was put in the oven at 18:33.
    What time must it come out of the oven?   ☐

4   Complete the missing times on this chart.

Flight	Departure time	Arrival time	Total journey time
TT391	06:54	08:33	
BEL203	09:36		1 hour 15 mins
JAM657		12:02	2 hours 2 mins
STL116	13:28		1 hour 40 mins
BAH209	19:47	23:16	

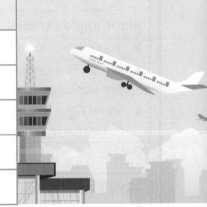

We write 12-hour time with a . between the hours and minutes, for example 11.30 a.m.

We write 24-hour time with a : between the hours and the minutes, for example 13:57.

Thinking mathematically

Grandad Billy uses a watch that loses 2 minutes each day.
At the start of the first day of each month he sets it to the correct time.

If you ask him for the time, he asks back, 'What is the date?'
He can then tell you the time.

a   If his watch says 11.20 a.m. on the 15th March,
    what is the actual time?   ☐

b   If the actual time is 3.45 p.m. on the 21st April,
    what time will Grandad Billy's watch show? ☐

Geometrical reasoning, shapes and measurements

6.1 Properties of polygons

Do you remember?

Polygons are straight-sided, closed shapes. If a polygon has equal sides and equal angles, it is **regular**. A polygon is **irregular** if it has sides and angles that are different.

These are the properties of different **triangles**:

Equilateral triangle	Isosceles triangle
• 3 equal sides • 3 equal angles 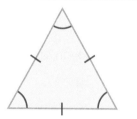	• 2 equal sides • 2 equal angles 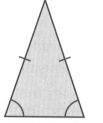
Right-angled triangle	**Scalene triangle**
• 1 angle is a right angle • some right-angled triangles are isosceles • some right-angled triangles are scalene 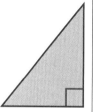	• no equal sides • no equal angles 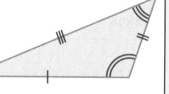

These are the properties of different **quadrilaterals**:

Square	Rectangle	Rhombus
• 4 equal sides • 4 right angles	• 2 pairs of equal sides • 4 right angles 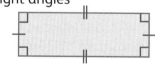	• 4 equal sides • opposite angles are equal • opposite sides are **parallel** 
Parallelogram	**Kite**	**Trapezium**
• opposite sides are equal and parallel 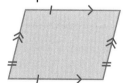	• 2 pairs of adjacent sides are equal	• 1 pair of parallel sides 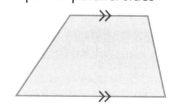

Maths words

polygon	equilateral triangle	quadrilateral	parallel
regular	isosceles triangle	square	parallelogram
irregular	right-angled triangle	rectangle	kite
triangle	scalene triangle	rhombus	trapezium

Try this

1 Which of these is a rhombus?

A 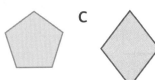 B C

2 Which of these shapes does not have two pairs of parallel sides?

A square

B rhombus

C regular pentagon

3 What is the name of this shape?

A kite

B trapezium

C tetrahedron

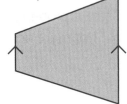

4 Which of these is a right-angled triangle?

A 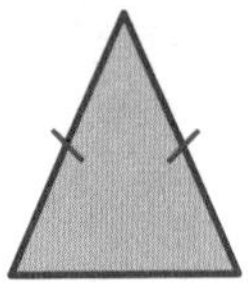 B 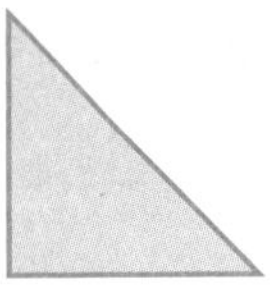 C

Practise

1 Complete this table of polygons.

Shape	Number of sides	Number of right angles	Number of pairs of parallel lines
equilateral triangle			
rhombus			
rectangle			
regular pentagon			
parallelogram			
right-angled triangle			
kite			

2 Complete these sentences by writing <u>always</u>, <u>sometimes</u> or <u>never</u>.

a A triangle _____ has a right angle.

b A hexagon is _____ symmetrical.

c A pentagon _____ has all sides the same length.

d A square is _____ symmetrical.

e A quadrilateral _____ tessellates.

f A parallelogram is _____ symmetrical.

g A rhombus _____ has all sides the same length.

h A parallelogram _____ has a right angle.

> Draw different forms of the shape each time. It will help you to see if the statement is always true, sometimes true or never true.

3   Draw one straight line between opposite corners of these shapes to make triangles.
    Describe the triangles. Use a ruler to check the length of the sides.

A

B

C

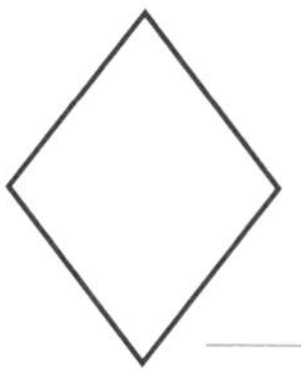

4   Draw four different quadrilaterals, each with a different property.

    a   equal sides   b   1 or more right angles

    c   parallel sides   d   opposite angles equal

Thinking mathematically

Two lines have been drawn in each rectangle to make new shapes.

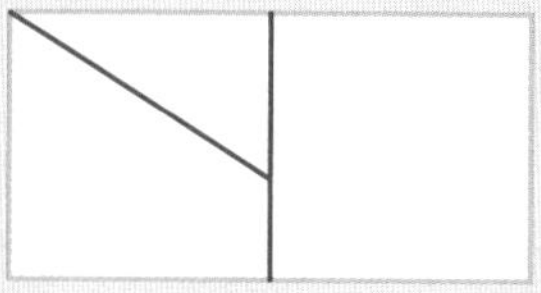

1 triangle
1 square
1 trapezium

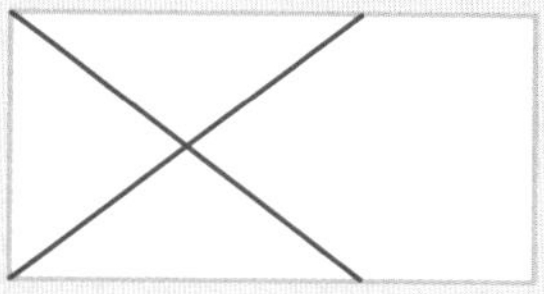

3 triangles
1 pentagon

Draw two lines in each rectangle to make the new shapes.

3 triangles

4 quadrilaterals

2 triangles
2 quadrilaterals

4 isosceles triangles

1 rectangle
2 right-angled triangles

2 trapeziums
2 right-angled triangles

6.2 Circles

Maths words
circle
radius
diameter
perimeter
circumference

Do you remember?

Look at the parts of a **circle**.

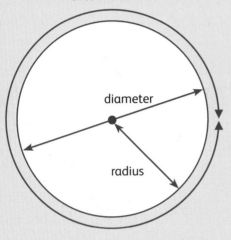

circumference

diameter

radius

The **radius** is the distance from the centre of the circle to the edge.

The **diameter** is the distance right across the circle through the centre. It is twice the length of the radius.

The **perimeter** of the circle is called the **circumference**. It is the distance all the way around.

We can use a pair of compasses to draw circles. The distance from the point to the pencil is the radius of the circle.

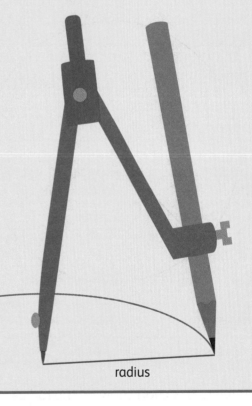

radius

Try this

1 What is the name of this part of a circle?

A radius

B circumference

C diameter

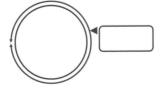

2 What is the name of this part of a circle?

A radius

B circumference

C diameter

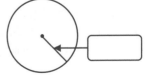

3 If the radius of a circle is 8.2 cm, what is the diameter?

A 4.1 cm   B 11.34 cm   C 16.4 cm

4 If the diameter of a circle is 19 cm, what is the radius?

A 38 cm   B 9.5 cm   C 6.5 cm

Practise

1 Measure the diameter and radius of each of these circles.

a
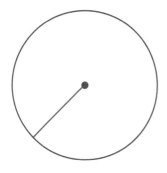

radius: ___ cm

diameter: ___ cm

b
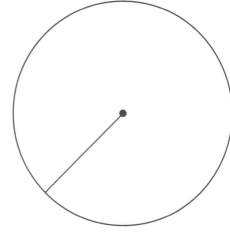

radius: ___ cm

diameter: ___ cm

c
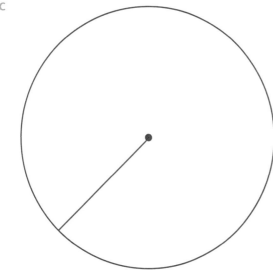

radius: ___ cm

diameter: ___ cm

d
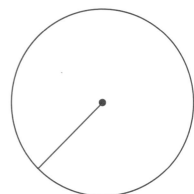

radius: ___ cm

diameter: ___ cm

2 Complete this chart showing the radius and diameter of different circles.

Radius	8 cm		6.5 cm		9.5 cm	
Diameter		18 cm		9 cm		21 cm

3 Josh thinks that he has drawn the diameter on this
circle because it goes from one side of the circle to the other.
Do you agree with Josh?
Explain your reasoning.

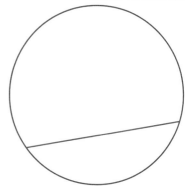

_____

_____

4   Use a pair of compasses to draw the following circles accurately.

   a   a circle with a radius of 4.5 cm

   b   a circle with a diameter of 11 cm

Thinking mathematically

Calculate the radius and the diameter of these circles.

a   The length of the line is 60 cm.

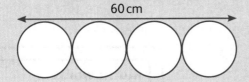

60 cm

b   The perimeter of the square is 60 cm.

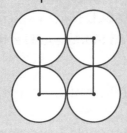

c   The perimeter of this equilateral triangle is 60 cm.

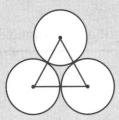

6.3 Perimeter and area of shapes

Do you remember?

The **area** of **compound shapes** made from **rectangles** can be found by working out the area of each part.

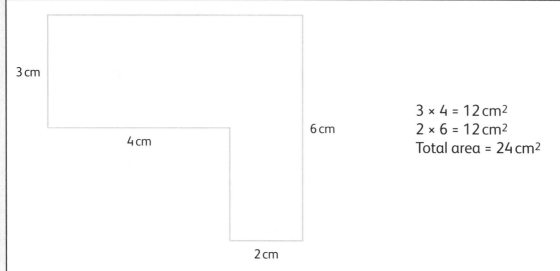

3 cm

4 cm

6 cm

2 cm

3 × 4 = 12 cm²
2 × 6 = 12 cm²
Total area = 24 cm²

The **perimeter** of the shape is the distance around the edge of that shape.

3 cm + 6 cm + 6 cm + 2 cm + 3 cm + 4 cm = 24 cm

Look at this **right-angled triangle**. It is half a rectangle.

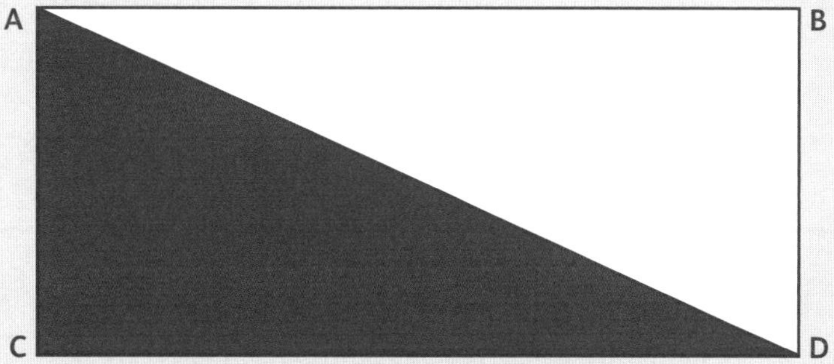

A

B

C

D

The area of this rectangle ABCD is 4 cm × 9 cm = 36 cm².

The area of the right-angled triangle ACD is half the area of the rectangle ABCD.

$\frac{1}{2}$ × 36 cm² = 18 cm².

The right-angled triangle ABD is also half the rectangle and has an area of 18 cm².

The area of a right-angled triangle = $\frac{1}{2}$ (area of the rectangle).

Maths words

area

compound shapes

rectangles

perimeter

right-angled triangle

Try this

1 A square has an area of 144 cm².
What is the length of each side?

   A 16 cm

   B 11 cm

   C 12 cm

3 Calculate the area of this shape.

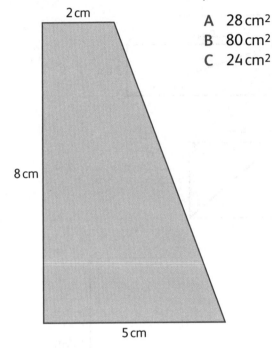

   A 28 cm²

   B 80 cm²

   C 24 cm²

2 What is the area of the red triangle?

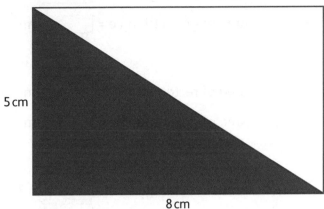

   A 40 cm²   **B** 10 cm²   **C** 20 cm²

4 A square has an area of 64 cm².
What is the perimeter of the square?

   A 32 cm

   B 64 cm

   C 24 cm

Practise

1 Calculate the area and perimeter of these letters.

  a Area = ☐ cm²

    Perimeter = ☐ cm

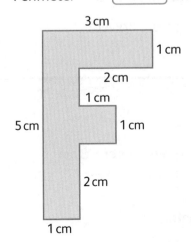

  b Area = ☐ cm²

    Perimeter = ☐ cm

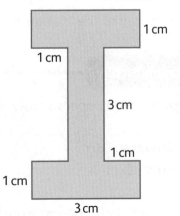

2   Calculate the area of each rectangle and the area of each green triangle.

a   Area of rectangle = ⬚ m²

Area of green triangle = ⬚ m²

8 cm

5 cm

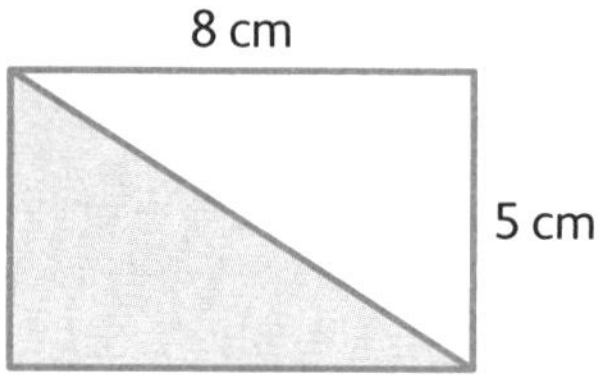

b   Area of rectangle = ⬚ m²

Area of green triangle = ⬚ m²

7 cm

4 cm

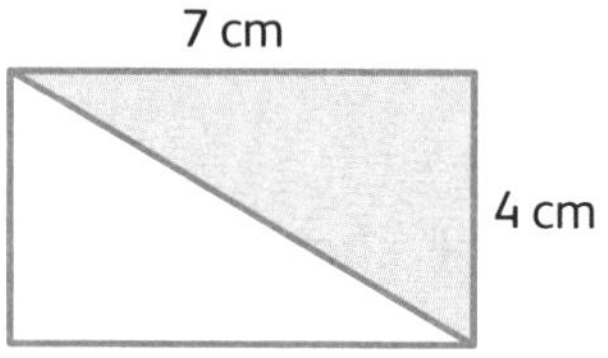

c   Area of rectangle = ⬚ m²

Area of green triangle = ⬚ m²

9 cm

3 cm

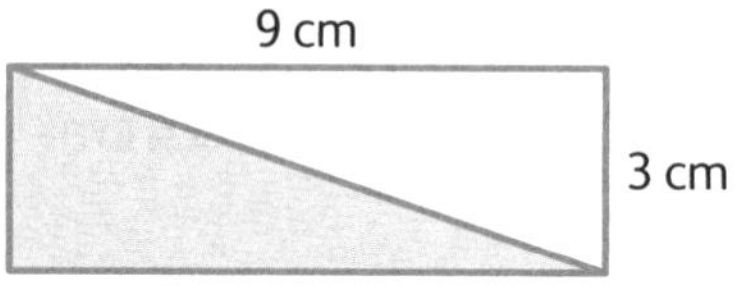

d   Area of rectangle = ⬚ m²

Area of green triangle = ⬚ m²

6 cm

5 cm

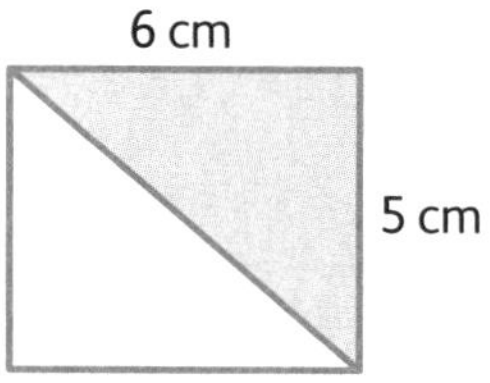

3   a   What is the area of the blue triangle? ⬚

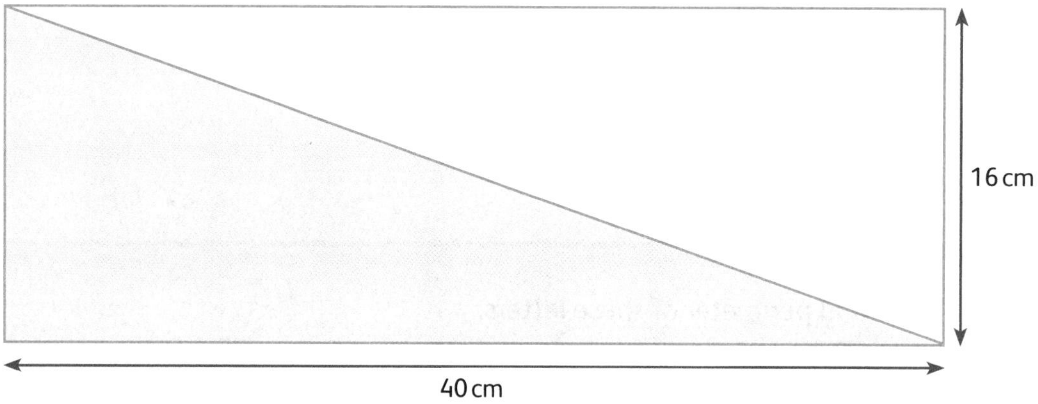

16 cm

40 cm

b   What is the area of the triangle if the sides of the rectangle are doubled? ⬚

c   What is the area of the triangle if the sides of the rectangle are halved? ⬚

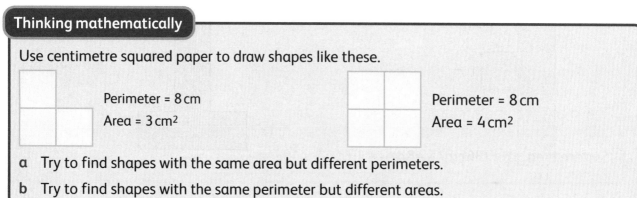

Thinking mathematically

Use centimetre squared paper to draw shapes like these.

Perimeter = 8 cm
Area = 3 cm²

Perimeter = 8 cm
Area = 4 cm²

a   Try to find shapes with the same area but different perimeters.

b   Try to find shapes with the same perimeter but different areas.

6.4   3D shapes

Do you remember?

Solid or 3D shapes are made up of **faces**, **edges** and **vertices** (corners).

A **cuboid** has 6 faces, 12 edges and 8 vertices.

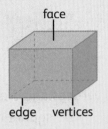

face

edge   vertices

Prisms

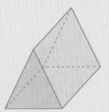

Triangular prism

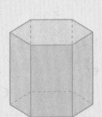

Hexagonal prism

Cuboids and cubes are special types of prism.

Pyramids

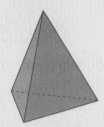

Triangular
pyramid

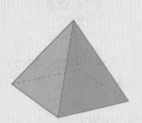

Square-based
pyramid

Another name for a triangular pyramid
is a tetrahedron.

The **net** of a shape is what it looks like when it is opened out flat.

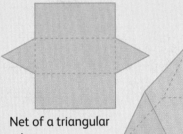

Net of a triangular
prism

To find the surface area of prisms you find the areas of its faces and add them together.

You can use nets of shapes to help work out surface areas.

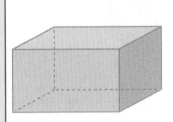

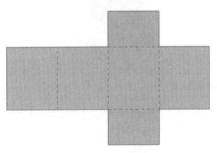

The area of the top face and bottom face = 12 cm × 3 cm = 36 cm²

The area of the front face and back face = 12 cm × 4 cm = 48 cm²

The area of one side face = 4 cm × 3 cm = 12 cm²

Surface area = 2 × (36 cm² + 48 cm² + 12 cm²) = 192 cm²

Try this

1 Which of these is a tetrahedron?

A

B

C

2 Which shape has 5 faces, 5 vertices and 8 edges?

A square-based pyramid   B square-based prism   C tetrahedron

3 Which of these has 6 vertices?

A

B

C

4 How many edges does a pentagonal pyramid have?

A 8   B 10   C 12

Practise

1 Write the name of the shape described in each of these.

A    B   C   D

a Which shape has 5 faces, 8 edges and 5 vertices?

b Which shape has 4 faces, 6 edges and 4 vertices?

c Which shape has 5 faces, 9 edges and 6 vertices?

d Which shape has 6 faces, 12 edges and 8 vertices?

2 Write the name of each of these shapes from its net.

a

b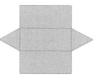

c

d

e

f

3   Write the name of each shape in the correct part of this Carroll diagram.

	No triangular faces	One or more triangular faces
No square or rectangular faces		
One or more square or rectangular faces		

4   Write the number of faces, edges and vertices on each shape in this chart.

Name of shape	Number of faces	Number of vertices	Number of edges
Tetrahedron			
Square-based pyramid			
Triangular prism			
Cuboid			
Pentagonal prism			
Hexagonal prism			

Look at the number of faces, edges and vertices for each shape on the chart.
Can you spot a rule or pattern between the numbers?

5   Calculate the surface areas of these cuboids.

a

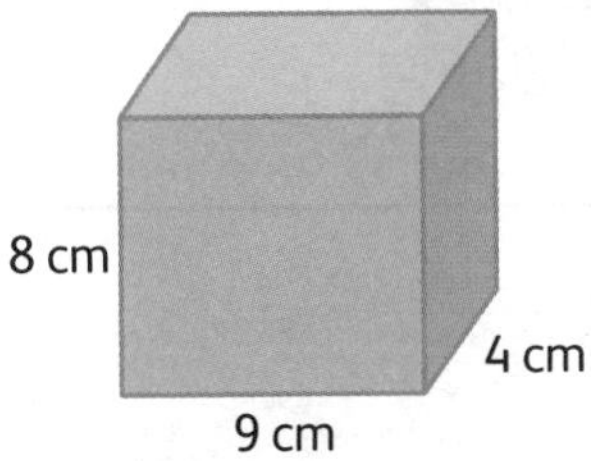

8 cm
4 cm
9 cm

b

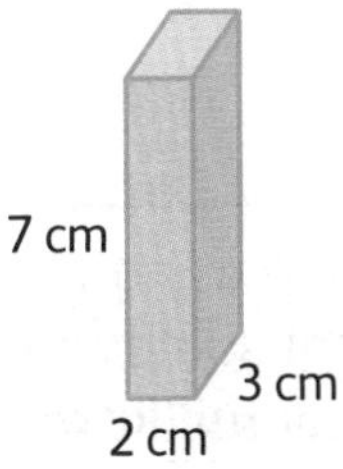

7 cm
3 cm
2 cm

Thinking mathematically

This shape is made from four cubes.
It is a **compound shape**.
Use cubes to make different compound shapes with four cubes.
Sketch them on isometric paper.

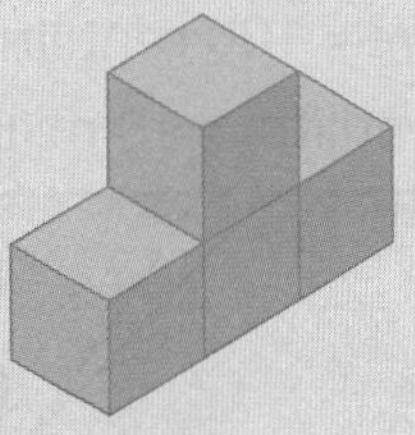

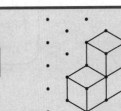

How many different shapes did you make?

How many compound shapes can you make with five cubes?

6.5 Capacity and volume

Do you remember?

Volume is the amount of space that something takes up.
Capacity is the amount that something will hold.

This jug has a capacity of 1 litre.
It is filled with 0.75 litres of water.
So, the volume of water in the jug is 0.75 litres.

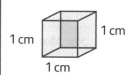

This cube is 1 cm long, 1 cm wide and 1 cm high.
The volume of the cube is 1 cubic centimetre or 1 cm³.
This is the amount of space that it takes up.

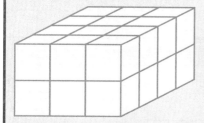

This cuboid has been made using 1 cm cubes.
It is 4 cm long, 3 cm wide and 2 cm high.
The volume is 24 cubic centimetres or 24 cm³.

Volume of a cuboid = length × width × height
Volume = *l* × *w* × *h*

Look at this cuboid:
Its dimensions are 10 cm × 3 cm × 4 cm.
Its volume is 120 cm³.

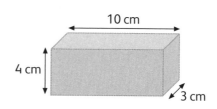

Capacity can be measured using different units.
The amount of liquid that something can hold is usually measured in
litres (ℓ), **centilitres (cl)** or **millilitres (ml)**.

1 litre = 1 000 millilitres
1 litre = 100 centilitres
1 centilitre = 10 millilitres

Maths words
volume
capacity
litre (ℓ)
centilitre (cl)
millilitre (ml)

Try this

1 What is the volume of this cube?

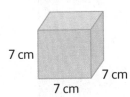

7 cm
7 cm
7 cm

A 28 cm³

B 343 cm³

C 283 cm³

2 What is the capacity of this jug?

3 litres
2 litres
1 litres

A 3ℓ

B 2.5ℓ

C 0.5ℓ

3 What is the volume of water shown in this jug?

1 litre
900 ml
800 ml
700 ml
600 ml
500 ml
400 ml
300 ml
200 ml
100 ml

A 45ℓ

B 0.45ℓ

C 0.4ℓ

4 What is the volume of this cuboid?

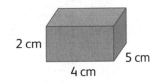

2 cm
4 cm
5 cm

A 40 cm³

B 11 cm³

C 245 cm³

Practise

1 Write the capacity and volume shown on these jugs.

a

1 litre
900 ml
800 ml
700 ml
600 ml
500 ml
400 ml
300 ml
200 ml
100 ml

Capacity = _____ litres

Volume of water = _____ litres

b

2 litres

1 litre

Capacity = _____ litres

Volume of water = _____ litres

c

3 litres

2 litres

1 litre

Capacity = _____ litres

Volume of water = _____ litres

2 Calculate the volume of these cuboids.

a

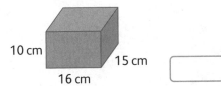

10 cm
15 cm
16 cm

b

50 cm
30 cm
90 cm

c

6.5 cm
2 cm
8 cm

d

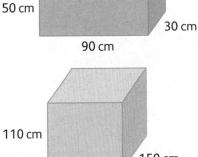

110 cm
150 cm
130 cm

3   The table shows the sizes of some cuboids.
    Some of the dimensions are missing. Complete the table.

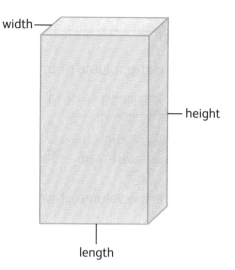
width
height
length

	Length	Width	Height	Volume
a	9 cm	6 cm	7 cm	
b		7 cm	10 cm	840 cm³
c	20 cm		12 cm	3 840 cm³
d	25 cm	20 cm		12 500 cm³
e	30 cm	10 cm	24 cm	

4   Complete the table.

	litres	centilitres	millilitres	cm³
a	2.5 ℓ			2 500 cm³
b			16 000 ml	
c		470 cl		
d			5 130 ml	
e	45.12 ℓ			

Thinking mathematically

What is the height of each of these fish tanks?

a   Fish tank has a volume of 3.75 litres.

Height of tank = ☐ cm

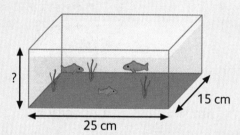
?
15 cm
25 cm

b   Fish tank has a volume of 10 m³.

Height of tank = ☐ cm

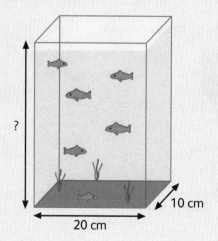
?
10 cm
20 cm

6.6   Reflective and rotational symmetry

Do you remember?

A **line of symmetry** divides a shape in half.
One half is the **reflection** of the other half.

The line of symmetry is the same as a **mirror line**.

Some shapes have no lines of symmetry; others have
one or more.

Maths words

line of symmetry

reflection

mirror line

rotational symmetry

order of rotational symmetry

centre of rotation

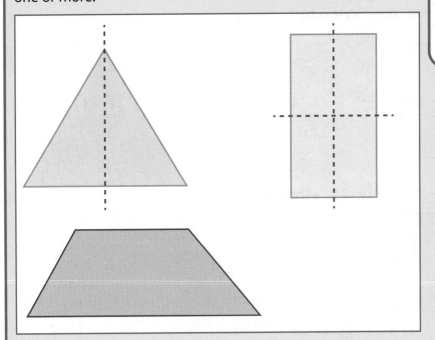

A shape has **rotational symmetry** if it fits on top of itself more than once as it takes a
complete turn.

The **order of rotational symmetry** is the number of times that the shape fits on top of itself.
This must be 2 or more. Shapes that only fit on themselves once have no rotational symmetry.

The **centre of rotation** (C) is the point about which the shape turns.

This shape has
rotational symmetry of
order 3 about its centre.

Try this

1   Which quadrilateral is symmetrical?

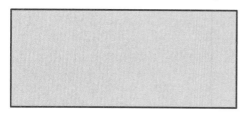

2   How many lines of symmetry does this shape have?

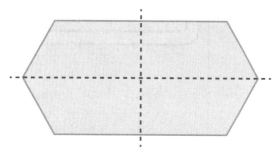

A 6   B 2   C 3

3   What is the order of rotational symmetry of this shape?

A 6   B 2   C 3

4   How many lines of symmetry does this shape have?

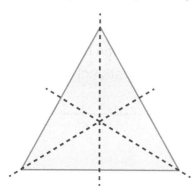

A 6   B 2   C 3

Practise

1   How many lines of symmetry do these triangles have?

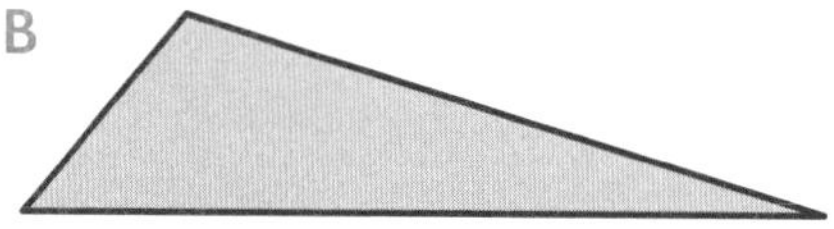

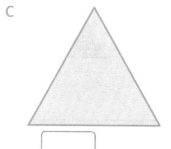

2   This is part of a tile pattern.

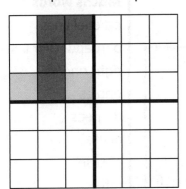

a   Colour squares to make the pattern have two lines of symmetry.

b   What is the order of rotational symmetry for your completed pattern?

3   Look at these symbols.

A   B   C   D

a   How many lines of symmetry are there for each symbol?

A [   ]   B [   ]   C [   ]   D [   ]

b   What is the order of rotational symmetry for each symbol?

A [   ]   B [   ]   C [   ]   D [   ]

4   Write the order of rotational symmetry for each shape.
Write 'none' if it has no rotational symmetry.

a 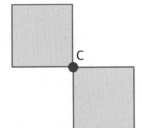   b 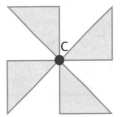   c 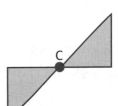   d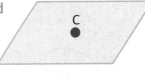

_____   _____   _____   _____

Thinking mathematically

This triangle has been rotated 180° around point C.
The triangle and its rotation make a parallelogram.

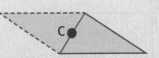

Copy each triangle. Give each triangle a 180° turn around C.
Write the name of each quadrilateral that you make.

a 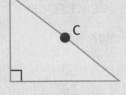   b 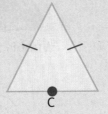   c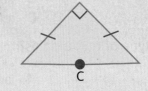

6.7 Angles, lines and shapes

Do you remember?

Parallel lines are the same distance apart from each other. Even if the lines are made longer, they will never meet.

Line AB is parallel to line CD.
Line AC is parallel to line BD.

Perpendicular lines are lines that meet or cross at **right angles**.

Angles ABC and ABD are both right angles. Line AB is perpendicular to line CD.

Angles are measured in **degrees** (°).
There are 360° in a full circle. These are special angles to remember:

180°
(straight line)

90°
(right angle)

acute angle
(less than a right angle)

obtuse angle
(between 90° and 180°)

reflex angle
(between 180° and 360°)

Angles on a straight line add up to 180°.

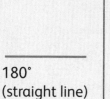

$a = 180° - 55° - 30°$
$a = 95°$

The angles of a triangle add up to 180°.

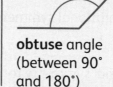

$a = 180° - 70° - 30°$
$a = 80°$

A protractor is used to measure the size of an angle.
It is a good idea to estimate the angle first and then measure it.

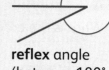

Try this

1 What type of angle is this?

   A acute angle
   B obtuse angle
   C reflex angle

2 What is the missing angle?

   A 65°
   B 55°
   C 125°

3 What type of angle is this?

   A acute angle
   B obtuse angle
   C right-angle

4 What is the missing angle?

   A 50°
   B 70°
   C 230°

Practise

1   Write the letter for each angle under the correct heading.

a   b 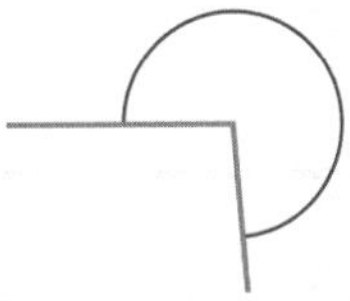   c 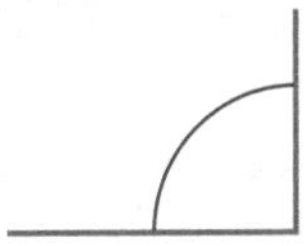   d

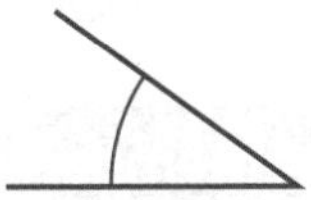

Acute angle	Right angle	Obtuse angle	Reflex angle

2   Now estimate, then measure each angle above and complete this chart.

Angle	a	b	c	d
Estimate (°)				
Measure (°)				

3   Write the size of the missing angle on each of these triangles.

a 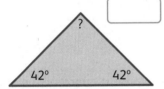   b    c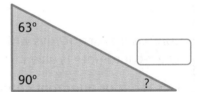

4   Write the missing angle for each of these.

a    b 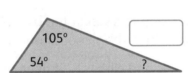   c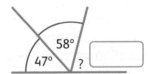

5   Look at this rectangle.

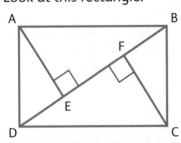

a   Use the letters to show the lines that are perpendicular.

_____

b   Use the letters to show the lines that are parallel.

_____

Thinking mathematically

Write the missing angles on these diagrams.

a 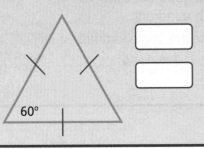   b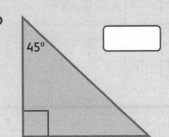

7.1 Read and plot coordinates

Maths words

coordinates

x-axis

y-axis

four quadrant

negative numbers

horizontal

vertical

origin

Do you remember?

Coordinates are used to show an exact position of a point on a grid. Two numbers from the **x-axis** and **y-axis** show the position:

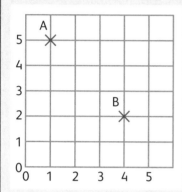

The coordinates of A are (1,5).
The coordinates of B are (4,2).

Coordinates are always written in brackets separated by a comma.
The x-coordinate is given first, and then the y-coordinate.

This is a **four quadrant** coordinates grid with **negative numbers** on the **horizontal** x-axis and **vertical** y-axis.

The point where the x-axis and the y-axis cross is called the **origin**.
The coordinates of this point are (0,0).
The x-axis and y-axis divide the space into four quadrants.

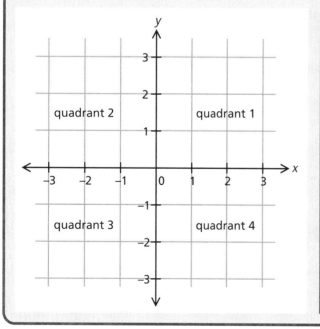

Shapes can be drawn on a coordinates grid.

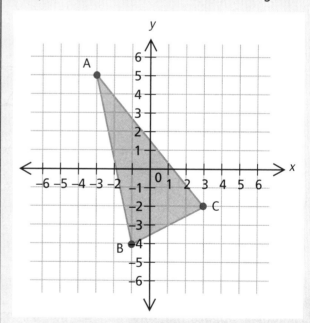

Point A is in the 2nd quadrant,
B in the 3rd quadrant and
C in the 4th quadrant.

The triangle has the coordinates:

A (–3, 5)   B (–1, –4)   C (3, –2)

Try this

1

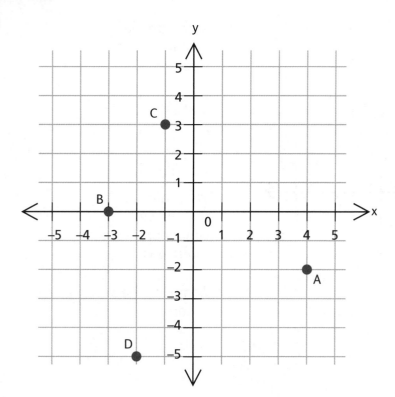

1 What are the coordinates for A?
   A (4, 2)
   B (4, −2)
   C (−2, 4)

2 What are the coordinates for B?
   A (0, −3)
   B (3, 0)
   C (−3, 0)

3 What are the coordinates for C?
   A (3, −1)
   B (−1, −3)
   C (−1, 3)

4 What are the coordinates for D?
   A (−5, −2)
   B (−2, −5)
   C (−2, 5)

Practise

1 Plot and label these points on the grid.

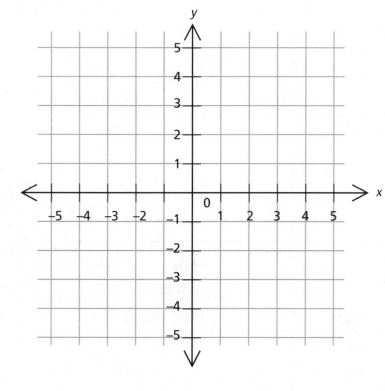

A (3, −5)
B (5, 0)
C (−3, −4)
D (−2, 5)

2   Here are two sides of a square.

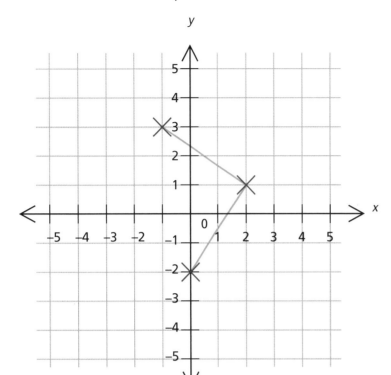

a   What are the coordinates of the three vertices?

(___ , ___),

(___ , ___),

(___ , ___),

b   What are the missing coordinates for the fourth vertex? (___ , ___)
Complete the square.

Thinking mathematically

What are the missing coordinates for vertex A and B?

A = (−40, ☐)

B = (☐ , 16)

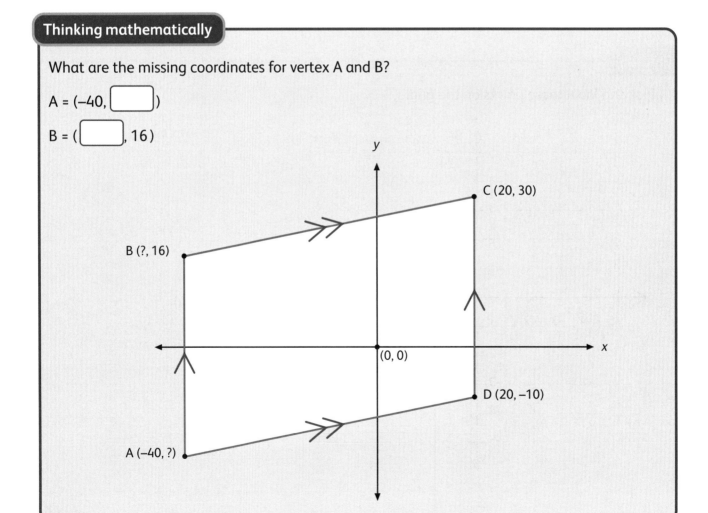

7.2 Transformation of shapes

A shape can be moved by **translation**, **reflection** or **rotation**.

Translation: sliding a shape without rotating or flipping it over. This shape has moved 4 squares across and 1 square down.

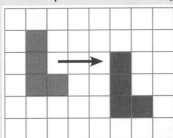

Maths words

translation
reflection
rotation
mirror line
clockwise
anticlockwise

Reflection: a shape is reflected or 'flipped' based on a **mirror line**.

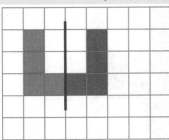

Rotation: a shape can be rotated about a point, **clockwise** or **anticlockwise**. Shape A is rotated clockwise around point X to become shape B.

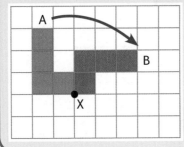

Try this

1 How has this shape been moved?

A   rotation

B   translation

C   reflection

2 How has this shape been moved?

A   rotation

B   translation

C   reflection

3 How has this shape been moved?

   A rotation

   B translation

   C reflection

4 How has this shape been moved?

   A rotation

   B translation

   C reflection

Practise

1 Write whether these shapes have been translated, rotated or reflected.

a

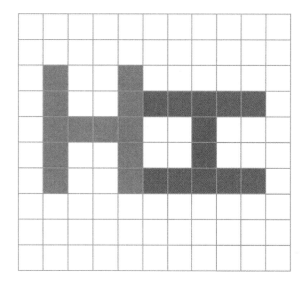

b

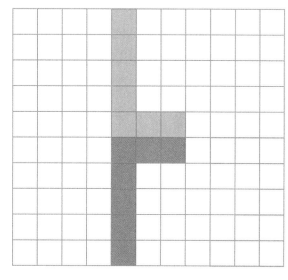

c

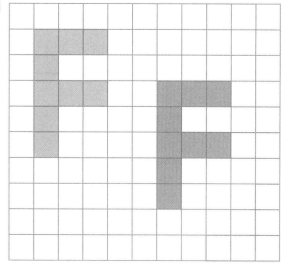

d

2   This triangle has been translated 5 squares across and 3 squares down.

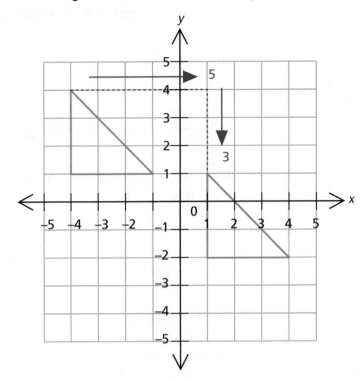

What are the coordinates of the
vertices of the translated triangle?

(___ , ___), (___ , ___), (___ , ___)

3   Plot these points on the grid and join them in order with a pencil and ruler:
   (4, 0),   (4, 1),   (1, 1),   (1, 4),   (0, 4)

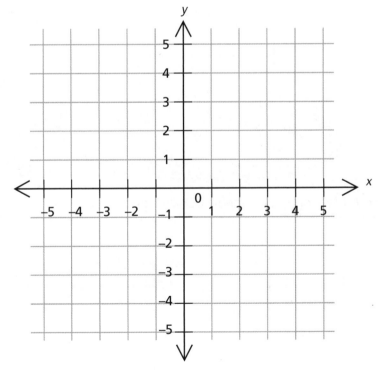

a   Reflect your drawing into the
   second quadrant.

b   Reflect both sets of lines in the
   x-axis. What shape have you
   made?

   _____

4 A triangle has been plotted on this grid.

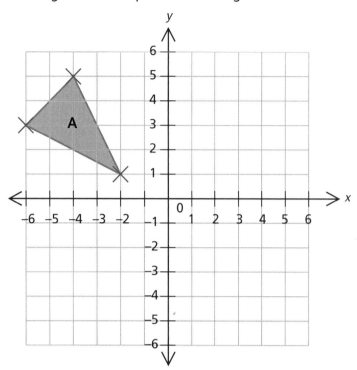

a What are the coordinates of triangle A?

(⬚ , ⬚),

(⬚ , ⬚),

(⬚ , ⬚)

b Draw triangle B at the following coordinates: (3, −3), (1, 1) and (−1, −1).

c Is triangle B a translation, rotation or reflection of triangle A?

_____

d Draw triangle C as a reflection of triangle A, with the y-axis as the mirror line.

e What are the coordinates of triangle C?

(⬚ , ⬚),

(⬚ , ⬚),

(⬚ , ⬚).

Thinking mathematically

This is your start tile.

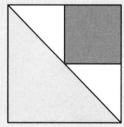

Choose to reflect, rotate or translate your tile. Copy and repeat it on this grid.

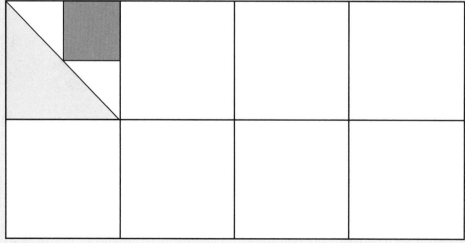

Design your own tile and explore the patterns you can make.

Test your understanding

1 a Draw hands on the clock to show 16.45.   **2 marks**
b What will the time be in 2 hours 35 minutes?

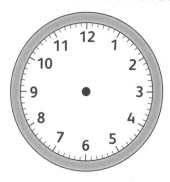

2 Here is a shaded square on a grid.   **1 mark**
Shade in 3 more squares so that the design is symmetrical in both mirror lines.

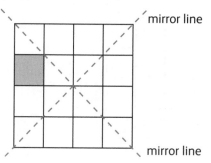

mirror line

mirror line

3   **3 marks**

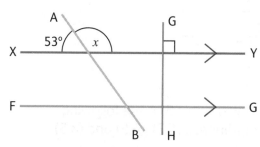

a Which line is perpendicular to XY?

b Which line is parallel to XY?

c What is the size of angle x?

4 Calculate the area and perimeter of this shape.   **2 marks**

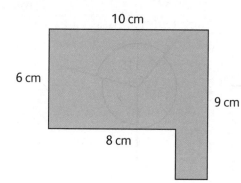

10 cm

6 cm

9 cm

8 cm

a Area =   cm²

b Perimeter =   cm

Test your understanding

5 Plot these points on the grid and join them in order with a pencil and ruler: **3 marks**
(5, 0), (1, 2) and (0, 5).

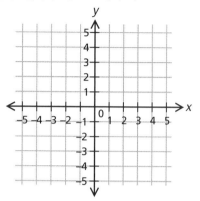

a Reflect your drawing in the second quadrant.

b Reflect both sets of lines in the *x*-axis. What shape have you made?

_____

6 Here is the net of a cube with no top. The shaded square is the base of the cube. Draw another square to show the missing top to complete the cube.   **1 mark**

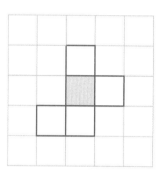

7 Triangle A is shown on a grid.   **3 marks**

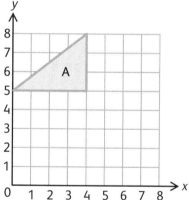

a What are the coordinates of triangle A?

(⬚ , ⬚),

(⬚ , ⬚),

(⬚ , ⬚)

b Draw triangle B at the following coordinates: (4,2), (4,5) and (8,5).

c Is triangle B a translation, rotation or reflection of triangle A?

_____

8 Calculate the value of angles *x*, *y* and *z*.   **3 marks**

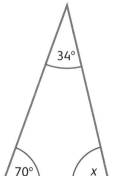

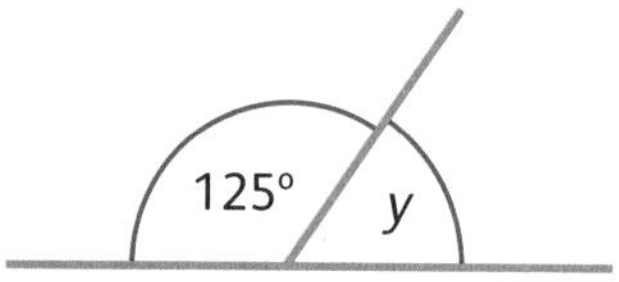

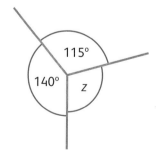

x = ⬚

y = ⬚

z = ⬚

Test your understanding

9 Triangle ABC is isosceles and has a perimeter of 30 cm.
Sides AB and AC are twice as long as BC.
What is the length of BC?

1 mark

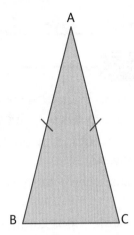

10 This shape is made from equilateral triangles and a square.
The square has sides of 15 cm.
What is the perimeter of the shape?

1 mark

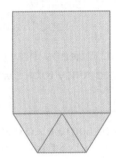

11 Write the name of each of these shapes from its net.

3 marks

a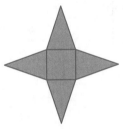

b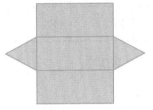

c

_____   _____   _____

12 Here are two sides of a square.

2 marks

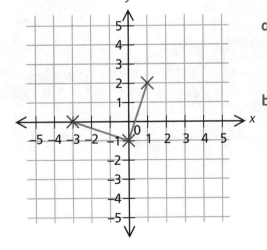

a What are the coordinates of the three vertices?

(___ , ___), (___ , ___), (___ , ___)

b Mark the missing coordinates for the fourth vertex to complete the square.

Total: _____ marks out of 25

8.1   Discrete data: bar charts and frequency tables

Do you remember?

Graphs and charts are a useful way of showing information. To understand **bar charts** and other types of graphs, look carefully at the different parts:

- Read the title – what is it all about? Is there any other information given?
- Look at the **axis** labels – these should explain the lines that go up and across.
- Work out the **scale** – look carefully at the numbers – do they go up in 1s, 5s, 10s …?
- Compare the bars – read them across to work out the amounts.

Maths words

bar chart

axis

scale

frequency table

grouped data

The word frequency means 'how many', so a **frequency table** is a record of how many there are in a group. Frequency tables with **grouped data** can be made into a bar chart analysing the results.

A class of children completed a sponsored walk.
The laps are grouped so they can be compared:

Laps completed	6–10	11–15	16–20	21–25	26–30
Number of children	2	7	17	18	25

This grouped data can then be shown on a graph:

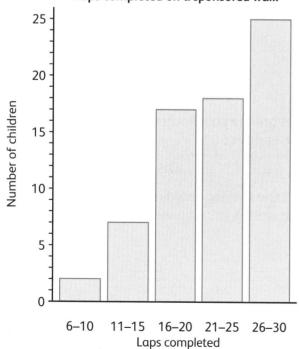

Laps completed on a sponsored walk

Try this

Look at the graph and table on the previous page to answer these questions.

1 Which number of laps did the largest number of children complete?

   A 11–15 laps
   B 21–25 laps
   C 26–30 laps

2 How many children completed 10 or fewer laps?

   A 9 children
   B 2 children
   C 7 children

3 How many children completed more than 20 laps?

   A 43 children
   B 25 children
   C 18 children

4 How many more children completed 16–20 laps than 11–15 laps?

   A 5 children
   B 15 children
   C 10 children

Practise

1 A kite shop recorded the sales of kites each day over two weeks.

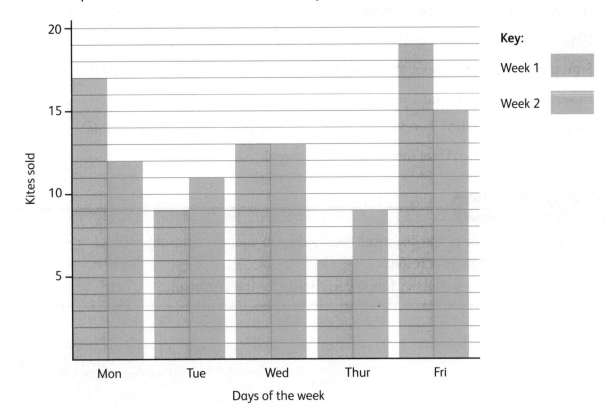

Key:
Week 1
Week 2

Look at the bar chart to answer these questions.

a How many kites were sold on Wednesday of Week 1?

b On which day of Week 2 was 11 kites sold?

c How many more kites were sold on Friday of Week 1 than Friday of Week 2?

d How many kites were sold in total on the two Mondays?

e What is the difference in sales between Week 1 and Week 2?

2 A survey was carried out to investigate the distances children travelled to school.

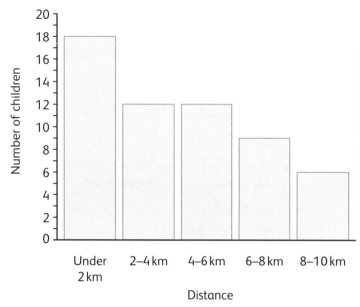

Use the graph to answer the questions.

a How many children travelled between 6–8 km to school?

b Which distance did the largest number of children travel to school?

c Sam travels 3.5 km to school. Which distance group would he be included in?

d How many children in total travel over 6 km to school?

e Half of the children travel under 4 km walk to school. How many children is this?

f How many children in total took part in this survey?

Thinking mathematically

Carry out a similar survey of the distances the children in your class travel to school.
Draw a bar graph to show your results.
Think carefully about the frequency range you choose for the distances.

8.2 Discrete data: pie charts and waffle diagrams

Pie charts are circles divided into **sectors**.
The whole circle represents 100 %. Each sector shows items as
a proportion so that they can be compared. The larger the angle,
the bigger the proportion. You can give the proportion as a fraction,
a percentage or a number.

Waffle diagrams show similar data to a pie chart but they are represented as a grid of 100
squares. The squares are coloured to show the proportions.

Maths words

pie charts

sectors

waffle diagrams

200 people were asked to choose their favourite type of film.

Pie chart

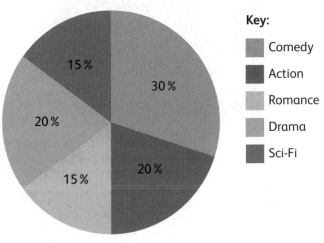

Key:

■ Comedy

■ Action

■ Romance

■ Drama

■ Sci-Fi

The largest proportion, 30 %,
chose comedy films.

15 % chose science fiction
films. 15 % of 200 is 30 people.

20 % of the group chose action
films. 20 % is $\frac{1}{5}$ as a fraction.

Waffle diagram

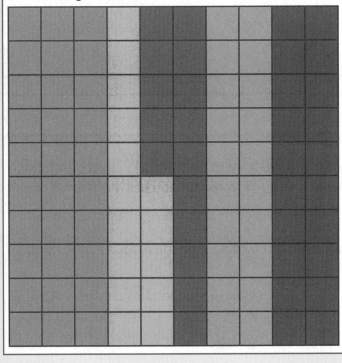

Do you remember?

When you compare the data from two pie charts look carefully at the totals for each, and the number of sectors.

These pie charts show the results of two basketball teams. Team A played 36 matches and Team B played 24 matches.

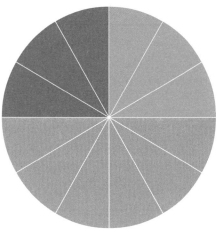

Team A match result

Team B match result

Key: ▢ Drawn   ▢ Won   ▢ Lost

Which team has won the most matches? It looks like the two teams have won the same number of matches but compare them carefully.

Team A won $\frac{1}{2}$ of 36 matches, which is 18.

Team B won $\frac{1}{2}$ of 24 matches, which is 12.

Try this

Use the pie charts of the basketball results above to answer these questions.

1 What percentage of matches did Team A lose?
    A 40%
    B 25%
    C 30%

2 How many matches did Team B draw?
    A 4
    B 6
    C 3

3 What fraction of matches did Team B lose?
    A $\frac{1}{3}$
    B $\frac{3}{10}$
    C $\frac{3}{8}$

4 How many matches did Team A draw?
    A 8
    B 12
    C 9

Practise

1   Ali went to a store to buy material to build a wall. The total cost was $220.
    This waffle chart shows how his money was spent.

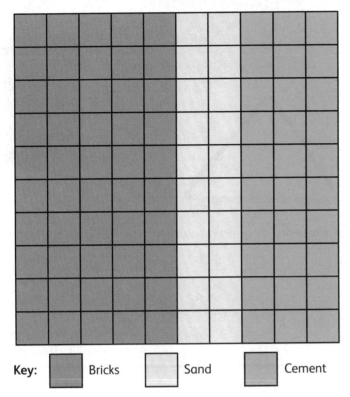

Key: ☐ Bricks   ☐ Sand   ☐ Cement

a   What fraction of his money
    was spent on cement?

b   How much did
    he spend on bricks?

c   What percentage of his
    money was spent on sand?

d   How much did
    he spend on sand?

e   How much did he
    spend on cement?

2   Ben also bought materials to build a wall. The total cost was $224.
    This pie chart shows how his money was spent.

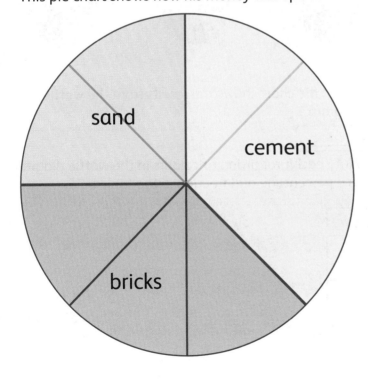

a   What fraction of his money
    was spent on bricks?

b   How much did he spend on sand?

c   What percentage of his money
    was spent on sand?

d   How much did he spend on bricks?

e   How much more did he spend on
    cement than Ali?

3   This pie chart shows the materials used by a class of 24 children to build model towers.
    Look at the pie chart and answer these questions.

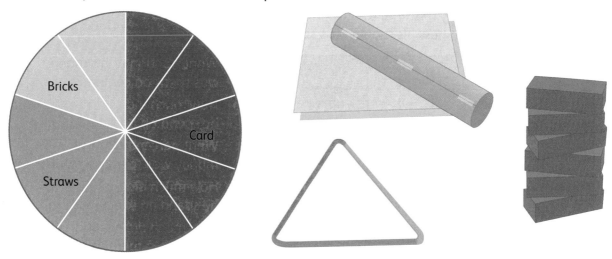

a   Which was the most popular choice of material to build a tower? _____

b   What fraction of the class used straws?

c   How many children used card?

d   What percentage of the class did **not** use bricks?

Thinking mathematically

Look at these diagrams.

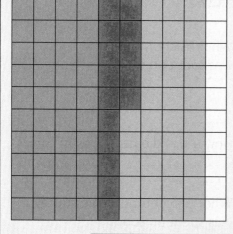

 A

 B

 C

a   Which pie chart shows the results from the waffle diagram?

b   Write each proportion of colours in the waffle diagram as a percentage and fraction in the table below.

	Percentage	Fraction
Blue		
Red		
Yellow		
Green		

8.3 Continuous data: line graphs and conversion charts

Maths words
line graph
continuous data
conversion chart
straight-line graph

Do you remember?

Line graphs have points plotted that are then joined with a line.

Line graphs are often used to represent measurements over time.

It is **continuous data**, so a reading can be taken from anywhere on the line.

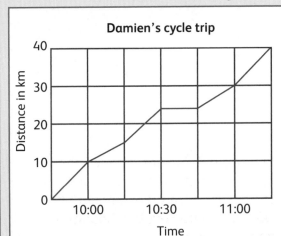

Damien's cycle trip

You can use the graph to show that Damien had cycled 30 km by 11.00.

Conversion charts are **straight-line graphs** that compare one unit of measure with another. They are also used to convert different currencies.

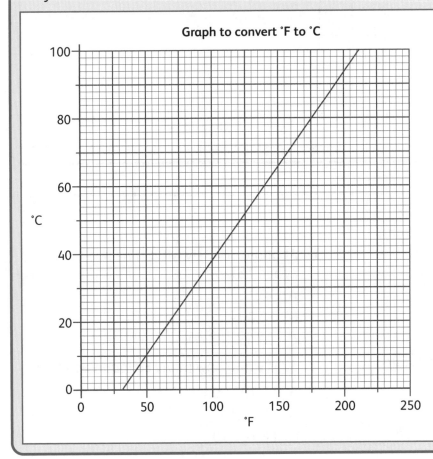

Graph to convert °F to °C

This graph converts from degrees Fahrenheit to Celsius.

You can use the graph to show that when it is 100°F it is approximately 38°C.

This table shows the conversion between US dollars ($) and Egyptian pounds (LE).

US dollars ($)	$1.00	$2.00	$3.00	$4.00	$5.00
Egyptian pounds (LE)	LE 15.30	LE 30.60	LE 45.90	LE 61.20	LE 76.50

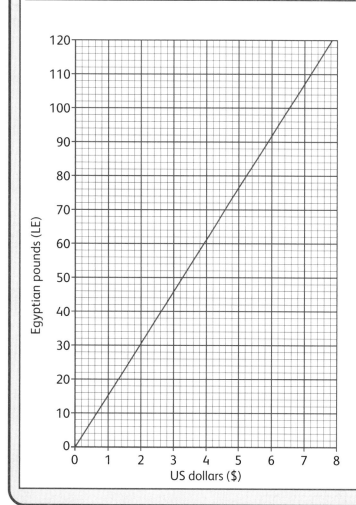

The straight-line graph shows the same information. We can use it to help us convert any amount of US dollars into Egyptian pounds and vice versa.

Try this

Use the graphs to answer these questions.

1 How far did Damien cycle between 10:00 and 11:00?

A 30 km

B 20 km

C 15 km

2 At approximately what time had Damien cycled 15 km?

A 10:15

B 10:30

C 10:45

3 What is the approximate temperature in Fahrenheit when it is 20°C?

A 70°F

B 90°F

C 50°F

4 $7 is worth how many Egyptian pounds?

A LE 102.10

B LE 97.10

C LE 107.10

Practise

1   Look at the graph and answer the questions.

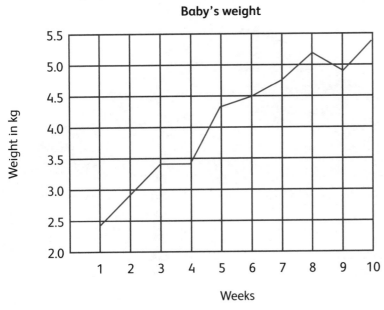

Baby's weight

a   When did the baby weigh 3 kg?

b   When did the baby weigh 4.5 kg?

c   When did the baby lose weight?

   Between ☐ and ☐ weeks

d   What was the approximate weight of the baby at 7 weeks?

e   During which time did the baby put on weight the fastest?

   Between ☐ and ☐ weeks

2   This graph shows the conversion rate between Euros and US dollars.

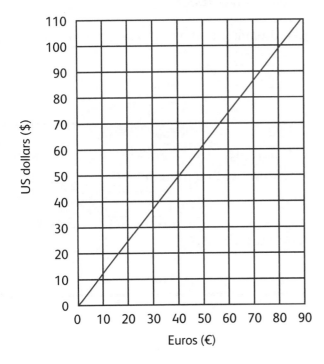

a   How many Euros can be exchanged for $50?

b   How many US dollars can I buy with €60?

c   How many US dollars can be exchanged for €20?

3   This graph shows the monthly average minimum and maximum temperature for Rome in Italy over a year.

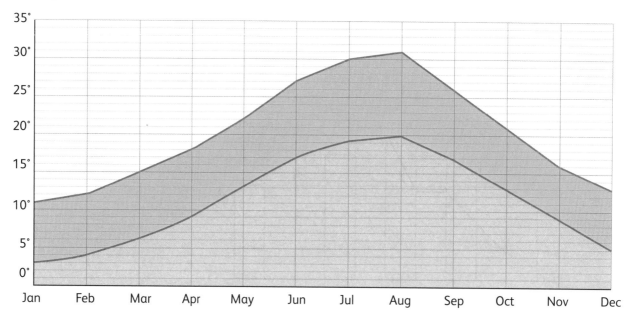

Key:

█ Max temp   █ Min temp

a   What was the average maximum temperature in May?

b   What was the hottest average maximum temperature?

c   In which month was the average minimum temperature 6˚C? _____

d   What was the difference between the hottest and coldest temperatures?

e   Which month had a maximum temperature 6˚C more than October? _____

f   Which month had the smallest difference between the minimum and maximum temperatures?

_____

Thinking mathematically

Water is leaking from a tap at a rate of 40 ml every 5 seconds.
a   Draw a line graph to show the rate of leaking water over a 20-minute period.

b   How much water is lost after 10 minutes?

c   How long will it take to fill a 1 litre jug, in minutes and seconds?

d   How much water is lost in 1 hour?

8.4 Mode, median, mean and range

Do you remember?

Maths words
mode
median
mean
average
range

Mode, **median** and **mean** are three types of **average**.

This chart shows the goals scored by the players in a football team.
What is the average for the number of goals scored?

Player	Al	Ben	Cara	Danni	Ed	Fred	Greta
Goals scored	7	5	3	8	6	3	3

Mode
The mode of a set of data is the value that occurs the most often. It is the most common number. Three players scored 3 goals so that is the mode.

Median
The median is the middle number when listed in order.
Put the numbers in order: 3, 3, 3, 5, 6, 7, 8
5 is the median number of goals.

Mean
The mean is what we normally think of as average.
Mean = total ÷ number of items.
Add the number of goals and divide by 7 as that is the number of players.
7 + 5 + 3 + 8 + 6 + 3 + 3 = 35
35 ÷ 7 = 5   So the mean average number of goals is 5.

The **range** is the difference between the highest and lowest values.
8 is the highest number of goals scored and 3 is the lowest. The range is 8 – 3 = 5.

Try this

1   What is the mode for this set of numbers?

A 14   B 17   C 18

2   What is the median for this set of numbers?

A 38   B 32   C 34

3   What is the mean for this set of numbers?

A 10   B 6   C 9

4   What is the range of this set of numbers?

A 40   B 35   C 25

Practise

1 This chart shows the number of children in each class in a school.

Class	1	2	3	4	5	6	7
Number of children	31	38	37	33	42	38	40

a Write the classes in order of size starting with the class with the smallest number of children.

Class							
Number of children							

b What is the median? Circle the median number of children.

c What is the mode? Colour the class sizes that show the mode.

d What is the mean number of children? Tick (✓) the class that has the mean.

e What is the range of class sizes?

2 These are the heights of a group of five children.

Ben: 140 cm	Sam: 130 cm	Eve: 140 cm	Amy: 150 cm	Jon: 190 cm

a What is the mode height?

b What is the median height?

c What is the mean height?

d How many children are above the mean average height?

e Another child joins the group. Her height is 120 cm.
   What is the mean height for the group now?

3 These are the hand-spans for a group of 10 children.

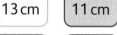

10 cm	8 cm
12 cm	9 cm
8 cm	10 cm
13 cm	11 cm
9 cm	10 cm

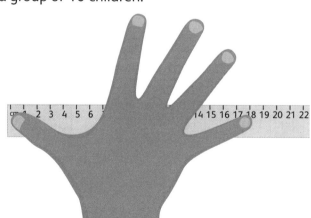

Complete these questions:

a Median: [   ]   b Mode: [   ]   c Mean: [   ]   d Range: [   ]

e What do you notice about the median, mode and mean?

_____

_____

Thinking mathematically

This is an information chart in a bike shop.

Name of bike	Gears	Frame size (cm)	Price
Big shot	1	29 cm	$125
Mustang	5	29 cm	$152
Activator	5	34 cm	$169
Calypso	1	29 cm	$115
Turbo	5	36 cm	$159
Chiltern	3	50 cm	$186
Pioneer	10	59 cm	$195
Sabre	18	50 cm	$179
Nitro	10	54 cm	$159
Ascender	15	48 cm	$165
Cassis	5	33 cm	$145

Use the chart to answer these.

a   What is the most common number of gears (the mode)?

b   Which bike has the median frame size?

c   What is the mean average price of these bikes?

d   Which bikes are at the mean average price?

BIKE SHOP

NEW

SALE

9.1   The language of probability

Do you remember?

We can use **probability** to predict what might happen in the future.
It is a measure of the **chance** or **likelihood** that it may happen.
This can be given as a fraction.

Even chance is an equal chance of something happening as not happening.
We also say a 1-in-2 chance or a 50:50 chance.

Look at this bag of beads.

What is the probability of
picking out a red bead?

There are 12 beads and
6 of them are red.
$\frac{6}{12}$ is the same as $\frac{1}{2}$,
so there is a 1-in-2
(or even) chance of
picking out a red bead.

What is the probability of
picking out a blue bead?

$\frac{2}{12}$ is the same as $\frac{1}{6}$,
so there is a 1-in-6 chance.
In theory this means that
for every 6 beads picked
out, 1 would be blue.

The probability scale
A probability scale uses language to show how likely an event is to happen:

| impossible | unlikely | even chance | likely | certain |

Where do you think these statements will be on the scale?

| I will fall asleep tonight. | I will break a world record this year. | Tomorrow will be Tuesday. | I can dive to the bottom of an ocean. |

Try this

1   When a 1–6 spinner is spun, what
    is the likelihood of it showing a 3?   A   1-in-3   B   1-in-5   C   1-in-6

2   What is the probability of throwing
    a coin and it landing to show heads?   A $\frac{1}{5}$ B $\frac{1}{2}$ C $\frac{1}{50}$

3   What is the likelihood of the sun
    rising tomorrow?   A   likely   B   certain   C   even chance

4   A bag holds 6 blue beads, 3 green
    beads and 9 red beads. A bead is
    picked out randomly. What is the
    chance that it is a red bead?   A   good chance   B   even chance   C   poor chance

Practise

1 Draw an arrow from each statement to the scale to show how likely each event is to happen.

| I will build a snowman tomorrow. | I will find $100 tomorrow. | I will eat bread tomorrow. |

| impossible | unlikely | even chance | likely | certain |

| Tomorrow will be Saturday. | It will rain tomorrow. | Tomorrow I will wear a hat. |

2 These ten number cards are shuffled and placed face down. Write a statement from the probability scale to show the likelihood of turning over these cards.

| 1 | 2 | 3 | 4 | 5 | 6 | 7 | 8 | 9 | 10 |

a a number greater than 6

_____

b a multiple of 2

_____

c a number that is not a 2-digit number

_____

d a multiple of 5

_____

e an odd number

_____

f the number 12

_____

3 Using a spinner, write the probability as a fraction of throwing each of these.

a an odd number

b a four

c a multiple of 4

d a seven

e a number greater than 2

f a number less than 3

4 On this spinner, what is the chance of spinning:

a red

b not red

c green

d orange

Thinking mathematically

Write something about yourself or an event to complete each of these probability sentences.

a There is a certain chance that _____

b It is impossible that _____

c There is a good chance that _____

d There is an even chance that _____

e There is a poor chance that _____

9.2 Probability experiments

Maths words
survey
prediction

Do you remember?

When predicting probabilities, it is sometimes useful to carry out a **survey** or experiment.
The larger the sample, the more reliable the **prediction**.

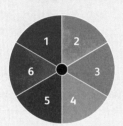

Spinners are useful for experiments and testing probabilities.
On a 1–6 (normal) spinner, the probability of throwing a 2 is a 1-in-6 chance or $\frac{1}{6}$.
What is the probability of throwing an even number?

You can use spinners for probability experiments.

- Cut out card squares and draw in diagonals.

- Colour the quarters in a pattern of your choice.

- Put a pencil through the centre so the spinner can spin.
- Predict the probability of landing on each of the colours on your spinner.

- Spin the spinner 100 times, recording the outcomes.
- Do the results agree with your predictions?

Try this with different spinners.

Try this

1 When a 1–6 spinner is spun, what is the likelihood of it showing an odd number?

   A 1-in-2 chance

   B 1-in-5 chance

   C 1-in-4 chance

2 A coin is tossed 20 times and lands heads up for the last 5 throws. What probability is there that it will land heads up on the 21st throw?

   A A good chance

   B A poor chance

   C An even chance

3   A bag holds 5 blue beads, 5 green beads and 10 red beads. A bead is picked out randomly. What is the chance that it is a green bead?

A $\frac{1}{5}$

B $\frac{1}{4}$

C $\frac{1}{2}$

4   This spinner is spun 100 times. How many times would you expect it to land on red?

A   20

B   25

C   50

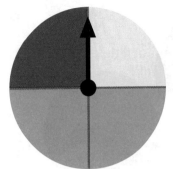

Practise

1   The beads in this bag are randomly picked out without looking. Write your answers as fractions.

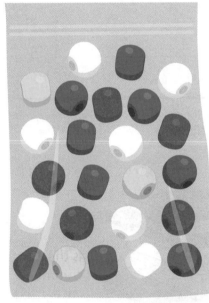

a   What is the probability of picking out a blue bead?

b   What is the probability of picking out a red bead?

c   What is the probability of not picking out a red bead?

d   What is the probability of picking out either a blue or yellow bead?

e   Sanjit randomly picked a bead out from this bag, replacing them each time. He repeated this 80 times. Approximately how many times would he be likely to pick out a yellow bead?

2   Answer this probability puzzle.

Which is most likely: a letter chosen at random from the alphabet being a vowel or spinning a six on a 1–6 spinner?

3   Circle the correct answer for these questions.

a   When a 1–6 spinner is spun, what is the likelihood of it showing a 4?

  A   1-in-3 chance

  B   1-in-4 chance

  C   1-in-5 chance

  D   1-in-6 chance

b   When two 1–6 spinners are spun, what is the likelihood of it showing a total of 12?

  A   1-in-12 chance

  B   1-in-18 chance

  C   1-in-24 chance

  D   1-in-36 chance

4 Carry out an experiment to investigate which totals are most likely to occur with two 1–6 spinners.

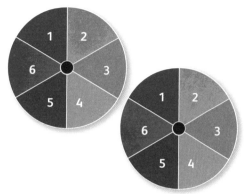

Instructions

- Spin two spinners.
- Add the numbers to make a total.
- Record your results on this table.

Total	Tally	Frequency
2		
3		
4		
5		
6		
7		
8		
9		
10		
11		
12		

Thinking mathematically

Choose one of the following statements or make your own prediction.
Check the predictions by carrying out a survey.

More than 1 in 2 people prefer eating chicken to beef.

The letter 'e' is the most common letter used in stories.

1 in 10 people are likely to be left-handed.

1 in 5 people choose the number 7 as their 'lucky' number.

Test your understanding

1 Anna has a part-time job in a shop.
She earns $8 for each hour that she works. This graph shows her wages.

2 marks

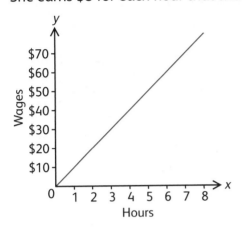

a How much would she earn for 3.5 hours work?

b How many hours would it take her to earn $56?

2 Look at these numbers.

2 marks

| 34 | 37 | 31 | 23 | 27 | 37 | 29 |

a What is the median of this set of numbers?

b What is the range of this set of numbers?

3 Shade this spinner so that there is a 1-in-3 chance that the arrow will land on a shaded part.

1 mark

4 These pie charts show the results of two football teams.
AFC played 16 matches and BFC played 18 matches.

3 marks

AFC match results

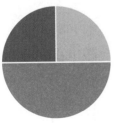

BFC match results

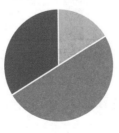

■ Drawn ■ Won ■ Lost

a Which team won the most matches?

b How many matches did AFC draw?

c What fraction of their matches did BFC lose?

Test your understanding

5 These are the results of a spelling test out of 10 for a group of 9 children: **3 marks**

| 8 | 6 | 7 | 7 | 8 | 5 | 5 | 8 | 9 |

a   What is the median score?

b   Which score is the mode?

c   Calculate the mean score.

6 A bag has 24 beads in it. 7 are red, 5 are green, 4 are blue and the rest are yellow.   **2 marks**

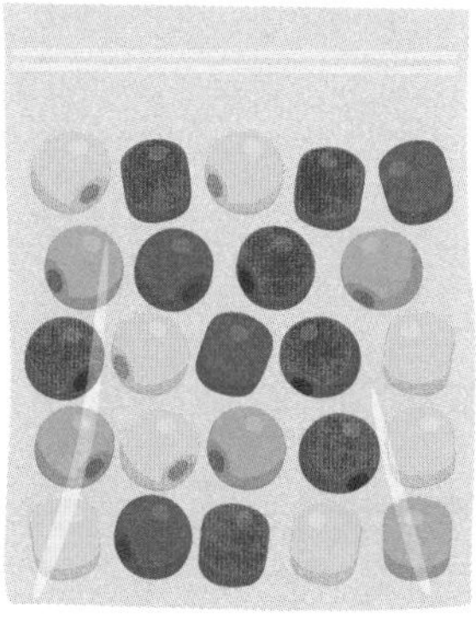

a   What is the probability of taking out a yellow bead?

_____

b   Another blue bead is added to the bag. What is the probability now of picking out a blue bead?

_____

7 A computer game has 30 levels of difficulty until it is completed. This graph shows the number of levels reached on the game by a group of children.   **3 marks**

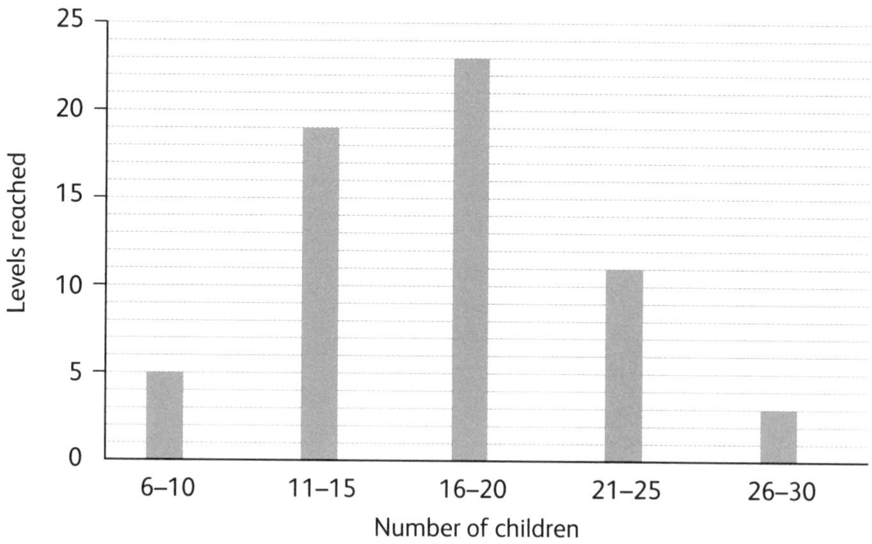

a   How many children reached between 21 and 25 levels?

b   How many children completed up to 15 levels?

c   How many children reached Level 21 or higher?

8 When two 1–6 spinners are spun, what is the likelihood of them showing a total of 2?   **1 mark**

A 1-in-12 chance   B 1-in-36 chance   C 1-in-24 chance   D 1-in-18 chance

_____   _____   _____   _____

Test your understanding

9 Mrs Jones set off at 08.00 and travelled 75 km in her car. **3 marks**

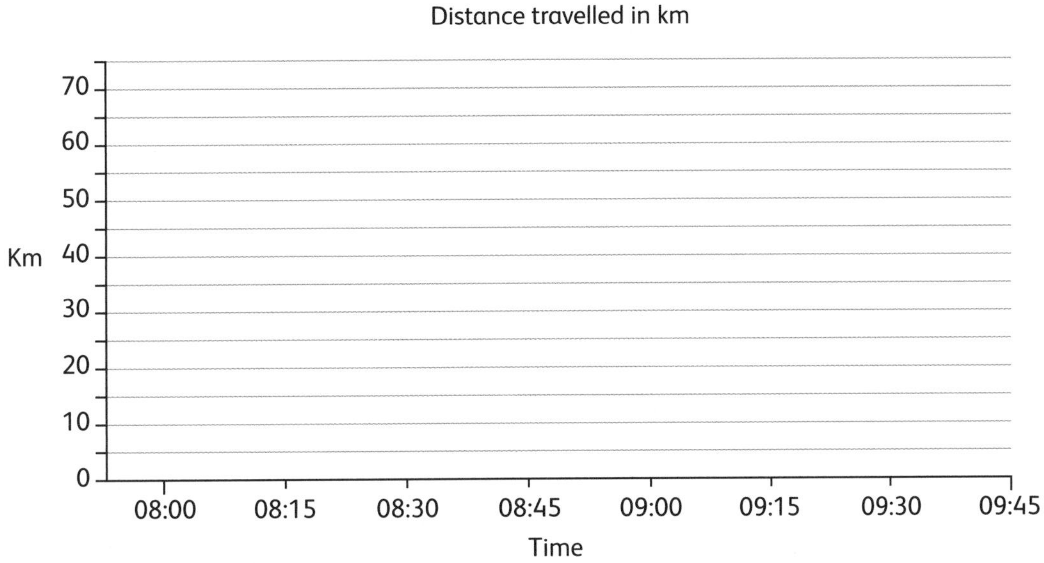

Distance travelled in km

a Draw a line graph using the information in the table.

Time	08:00	08:15	08:30	08:45	09:00	09:15	09:30	09:45
Distance (km)	0	10	25	30	30	45	60	75

b How far had she travelled after 30 minutes?

c At what time had she travelled one-third of the distance?

10 There are two possible results when you toss one coin: H and T. How many possibilities are there when you toss two coins? **1 mark**

11 These are the results of a maths test out of 100. **3 marks**

71   82   76   90   63   83   76   89   92   82   76

a Write the scores in order, starting with the lowest.

b Which score is the mode?

c Which score is the median?

12 On this spinner, what is the likelihood of the arrow pointing to either red or green? **1 mark**

_____

_____

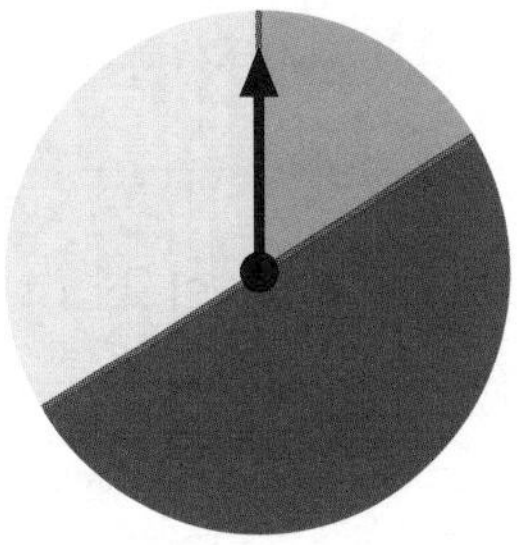

Total: _____ marks out of 25

109

My study notes

Glossary

acute angle – an angle between 0° and 90°

anticlockwise – turning in this direction

approximate – a 'rough' answer – near to the real answer

area – the area of a shape is the amount of surface that it covers

axis – (plural is axes) the horizontal and vertical lines on a graph

bar chart – a chart that uses bars of equal width to represent data

circle – a 2D shape with every point on its edge a fixed distance from its centre

circumference – the edge or perimeter of a circle

circumference

clockwise – turning in this direction

continuous data – continuous data can take any value and is arranged in groups

denominator – bottom number of a fraction, the number of parts it is divided into. Example: $\frac{2}{3}$

digits – there are 10 digits: 0 1 2 3 4 5 6 7 8 and 9 that make all the numbers we use

divisor – a divisor is a number that another number is divided by. For 32 ÷ 4 = 8 the divisor is 4

edge – where two faces of a solid shape meet

edge

equation – where symbols or letters are used instead of numbers. Example: 3y = 12, so y = 4

equivalent – two numbers or measures are equivalent if they are the same or equal

equivalent fractions – these are equal fractions. Example: $\frac{1}{2} = \frac{2}{4} = \frac{3}{6}$

even chance – if an event has an even chance there is the same chance of it happening as not happening

faces – the flat sides of a solid shape
face

factor – a number that will divide exactly into other numbers. Example: 5 is a factor of 20

formula – a formula (plural is formulae) uses letters or words to give a rule

frequency – the number of times that something happens is called the frequency

horizontal – a horizontal line is a straight level line across, in the same direction as the horizon

improper fraction – a fraction which has a numerator greater than the denominator

mean – this is the total divided by the number of items. The mean of 3, 1, 6 and 2 is (3 + 1 + 6 + 2) ÷ 4 = 3.

median – the middle number in an ordered list. Example: 3, 8, 11, 15, 16. The median number is 11.

mixed number – a whole number together with a proper fraction

mode – the most common number in a list. Example: 2, 6, 4, 2, 5, 5, 2. The mode is 2.

multiple – a multiple is a number made by multiplying together two other numbers

negative number – a number less than zero on the number line

net – the net of a 3D shape is what it looks like when it is opened out flat

numerator – is the top number of a fraction. Example: $\frac{3}{5}$

obtuse angle – an angle between 90° and 180°

origin – the point where the x-axis and y-axis cross, with coordinates (0, 0)

parallel – lines that are parallel never meet

percentage – this is a fraction out of 100, shown with a % sign

perimeter – the boundary or edge of an area

perpendicular – a perpendicular line is one that is at right angles to another line

pie chart – a circular chart illustrating data

polygon – any straight sided flat shape

prime factor – any factor that is a prime number is a prime factor

prime number – these only have two factors, 1 and itself. For example, 23 is a prime number as it can only be divided exactly by 1 and 23.

prism – a 3D shape with a constant cross-section

probability – the chance an event may happen, often given as a fraction

proportion – this is the same as finding the fraction of the whole amount. Example: the proportion of red cubes is 3 out of 5 or $\frac{3}{5}$

protractor – a tool for measuring angles

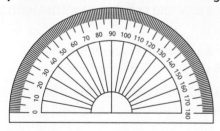

quotient – this is the number of times that one number will divide into another number. Example: When you divide 18 by 3 the quotient is 6

radius – distance from the centre to the circumference of a circle

radius

range – the difference between the greatest and least values of a set of data

ratio – this compares one amount with another. Example: the ratio of red cubes to blue cubes is 3:2.

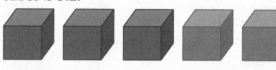

remainder – if a number cannot be divided exactly by another number then there is a whole number answer with an amount left over, called a remainder

rotation – when a shape is rotated, it is turned around a centre of rotation and through a given angle, either clockwise or anti-clockwise

sequence – a list of numbers which usually have a pattern. They are often numbers written in order.

square number – numbers multiplied by themselves make square numbers
Example 4 x 4 = 16. The first five square numbers are 1, 4, 9, 16 and 25.

symmetrical – when two halves of a shape or pattern are identical

translation – a transformation in which every point of a shape moves the same distance and direction

vertical – a line that is straight up or down, at right angles to a horizontal line

vertices – (single is vertex) these are the corners of 3D shapes, where edges meet

vertex

volume – the amount of space in a 3D shape